Coal in Our Veins

Coal in Our Veins

A Personal Journey

Erin Ann Thomas

Utah State University Press
Logan, Utah
2012

Published by Utah State University Press
An imprint of University Press of Colorado
5589 Arapahoe Avenue, Suite 206C
Boulder, Colorado 80303

The University Press of Colorado is a proud member of

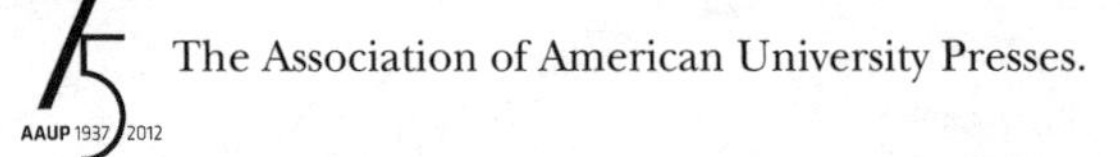

The Association of American University Presses.

The University Press of Colorado is a cooperative publishing enterprise supported, in part, by Adams State College, Colorado State University, Fort Lewis College, Metropolitan State College of Denver, Regis University, University of Colorado, University of Northern Colorado, Utah State University, and Western State College of Colorado.

Manufactured in the United States of America
Printed on recycled, acid-free paper
Cover design by Jeanne Goldberg

ISBN: 978-0-87421-863-3 (cloth)
ISBN: 978-0-87421-865-7 (e-book)

Library of Congress Cataloging-in-Publication Data
Thomas, Erin Ann.
Coal in our veins : a personal journey / Erin Ann Thomas.
p. cm.
Includes bibliographical references.
ISBN 978-0-87421-863-3 (cloth : acid-free paper) — ISBN 978-0-87421-865-7 (e-book)
1. Thomas, Erin Ann—Family. 2. Coal miners—Utah—Biography. 3. Coal miners—Wales—Biography. 4. Welsh Americans—Utah—Biography. 5. Thomas, Erin Ann—Travel. 6. Coal mines and mining—Utah—History. 7. Coal mines and mining—Wales—History. 8. Coal—Environmental aspects—United States. 9. Coal—Social aspects—United States. I. Title.
HD8039.M62U669417 2012
622'.3340922—dc23

2012012249

Dedicated to Robert K. Thomas, my grandfather,
coal miner, and educator

We all know that we must have coal, but we seldom or never remember what coal getting involves.... We are capable of forgetting ... as we forget the blood in our veins.

—George Orwell

Contents

Washington, D.C.

Coda

Introduction

Erin Thomas 2012

A carbide miner's lamp.

1

A Miner's Lamp

A carbide lamp, no longer used as a functional object, sits on my bookshelf in the living room next to other mementos, relics, and souvenirs from around the world. These objects remind me of the experiences I had in the cities where I purchased them, but displayed in my living room, they bring more attention to me than they do to the location of their origin. For my living room visitors, the carbide miner's lamp is a visual affirmation of my physical and literary journey into coal. Sometime between the years 1900 and 1930, a miner used this lamp to illuminate his way through the tunnels of a mine.

Carbide lamps were developed after candles and oil lamps, each an attempt to provide light for mining by a more efficient and safer means. My carbide lamp is about four inches tall and cylindrical—approximately two inches in diameter. It has two chambers: one that can be filled from the top with water and a lower chamber for carbide. This lamp is designed so that water from the upper chamber drips down into the lower, creating acetylene gas when it comes into contact with the whitish carbon "salt." These two chambers screw together like a light bulb into a socket.

When I unscrew my carbide lamp it squeaks slightly; in the chamber below a few crumbs of carbide remain—ashes left by somebody who passed on long ago. After placing the carbide in the lamp and screwing it shut, this miner would have opened a small spout in the middle of a metal disk-reflector in the front. This released the acetylene, which he would have ignited by sparking flint. He would then have turned the spout until the flame was about an inch tall, the reflector casting the light of the flame forward ten feet in a thick beam. This model, made by one of the most popular suppliers of the day, Justrite, has a rubber grip at the bottom of the brass body. The miner might have held it in his hand or hooked it to the top of his cloth hat. But the crumbs of carbide left in my lamp give no indication of the particulars of this miner's life or how the lamp made its way from his possession as much as a hundred years ago to an online antique store where my father purchased it for my twenty-ninth birthday.

In the United States, miners no longer used this sort of lamp by the time my grandpa, Robert Thomas, began mining coal alongside his father in the 1930s. He would have been called "Bobby" then and used a battery-powered light that provided less of a fire hazard in a gaseous mine. Bobby is my primary connection to coal, and growing up I was always conscious that my surname had been passed down through generations of strong-backed and rough-handed men. I have always been proud of this fact, preferring my ancestry of Welsh coal miners over a lineage of kings.

Despite this, the coal-mining stories my grandparents and parents told me were like relics or souvenirs. Through them, the memories of my grandfather, his parents, and those before them were preserved, but ultimately I regarded these stories as something that made me more interesting, rather than linking me to the reality of my ancestors' lives or even the reality of my grandfather's life. During the years I knew him, he was a respected educator in the world I lived in. As a former academic vice president at Brigham Young University (BYU) and the founder of its Honors Program, Bob Thomas was still quoted when I started school there in 1996. This, rather than coal, was the tradition that he passed down to my father, Ryan, who is also a college administrator. My grandfather's choice to leave mining redefined what it meant to be a Thomas. And as a Thomas and my father's daughter, I have also found my place in education, carving out my niche in teaching ESL and English, my occupation now in the Washington, D.C. metropolitan area. Until very recently, Bobby, a teenager in overalls with a blackened countenance, felt as distant as he was different.

This changed early in 2006 when I was walking through O'Hare Airport during a layover on my flight to D.C., returning from Christmas vacation. Hours before, I had been dropped off at the Salt Lake airport by my parents, who, after hugging me and placing my luggage on the sidewalk, traveled back to central Utah. They headed through the valleys and up a canyon to their home in Price, a former mining community of ten thousand residents, where my father was the president of the College of Eastern Utah. When he accepted the job five years previously, leaving his position at Utah Valley State College, I joked with my friends that he and my mother were moving south so he could become a coal miner. Price is near Castle Gate, where my grandfather grew up and my great-grandfather mined. This connection to our family history heavily influenced my father's decision to take the position.

My journey to D.C. symbolically leaped in tandem with my parent's journey back to Carbon County and its abandoned coal mines when the voice of a television reporter reached me over the noise of the terminal.

I peered up at a screen rigged over the departure gate ahead. The camera shifted from a newsroom to groups of people in T-shirts and jeans with strained looks on their faces in front of the backdrop of a little white church: thirteen miners had been trapped in a West Virginian coal mine.

In my almost thirty years there had been a number of mining accidents. During this time, if I had read a newspaper article or watched a news briefing of a mining explosion, I had never taken note of it. For most of my life, coal-mining stories had been exclusively family stories. But on this day these people in the small town of Sago, West Virginia, brought the past of my family jarringly to my present.

Two of my great-great-uncles met their deaths early in mines, and Robert's father, Zephaniah Thomas, barely escaped two of the worst mining disasters in US history: the Winter Quarters explosion of 1900 and the Castle Gate explosion of 1924, both in central Utah. Minutes after the Castle Gate explosion, Bobby, who was then five years old, stood on the porch and watched screaming women pour into the streets. He vividly remembered the 172 white coffins laid out on rough tables in the local amusement hall. They would have loomed large from a child's view below; only by chance did his family have no body to collect there.

In the O'Hare Airport, I was startled by the ordinariness of the families who gathered for news of their loved ones—coal mining moved from the legendary quality family stories acquire as they are passed down to the actual. For the first time I fully realized, *I am a coal miner's granddaughter.* As a result of this fact, some sort of responsibility bound me to these people.

The explosion occurred in a small mine in the town of Sago in Upshur County. A methane gas leak had somehow ignited, filling the mine with smoke and lethal fumes and blocking the escape route of a mining crew two and a half miles underground. At the time of the news report I saw in the airport, high carbon monoxide levels prevented rescuers from entering the mine. Teams stationed at the mouth of the mine tested gas levels periodically. Family and community members at the local Sago Baptist church waited for news of the thirteen miners trapped inside. Media vans clogged the narrow streets that crossed in front of the small chapel, and news cameras watched as locals held each other and cried—the intimate expressions of their distress broadcast to a screen in the terminal where I sat waiting for my flight to the District of Columbia. I lingered until it was near my departure.

In my bags, I carried a binder of genealogy assembled by my uncle and aunt for the family for Christmas. It burst at the rings with accounts of my mining ancestors through three generations, from their lives in central Utah back to the dawning of the Industrial Revolution in Merthyr Tydfil, Wales. Not long after watching the news report in O'Hare Airport, I pulled this binder off my bookshelf one evening. As the sun set and the night progressed into morning, I flipped through pages of transcribed oral histories, written personal histories, funeral speeches, and school compositions on parents and grandparents. Among these accounts were photocopies of old pictures, and I scrutinized each of their faces for a hint of my own. I had heard these stories since I was a child, but reading them that night in the context of a modern mining disaster was riveting in a way that the stories my parents told me never were. Zephaniah and Maud, and Evan and Margaret began to exist in their own right, not just as great-grandparents and great-great-grandparents who died before I knew them. They had struggled in the collier lifestyle, always wanting something different for themselves and their children. Each one of them had lost a family member in a mining accident—in cave-ins and explosions—just as miners die today.

Although the Sago mining accident was one of the most closely followed news stories of 2006, coal mining is something we rarely hear about except in times of extremity. Occasionally a social activist attempts to bring the conditions of the coal mine to light. Back in the days when my grandfather mined coal, George Orwell completed an investigative survey of the mines in northern England in 1936, using his experiences as the introduction to *The Road to Wigan Pier*, a book dedicated to exposing the conditions of the working class. He is not the most recent critic of an industry that has significant human and environmental costs, but one of the more eloquent and perceptive. I quote him because the first time I read his words, I felt they expressed something I was just beginning to articulate. He asserted: "Down there where the coal is dug it is a sort of world apart which one can quite easily go through life without ever hearing about." By lamplight, he watched men with blackened hands and faces extract a resource that by now has been the main source of carbon power—and the dirtiest—for two hundred years, fueling the technological revolution that has mechanized our society from the horse to the engine. The lifestyle that produced this energy was, as Orwell pointed out seventy-one years ago, the absolutely necessary counterpart of our world above: "Practically everything we do, from eating an ice to crossing the Atlantic, and from baking a loaf to writing a novel, involves the use of coal, directly or indirectly." This is almost as true now as it was then. Despite the variety of alternate forms of energy available today, the United States depends on coal-burning electric plants for 50 percent of the nation's power, and it is the second-largest producer of coal in the world.

In 1936, when George Orwell traveled to the northern areas of the United Kingdom to document poverty in the working class, Bob's father, Zephaniah Thomas, had moved his family from Castle Gate, Utah, to Oregon, where he unsuccessfully worked a number of small mines—Bobby beside him after school and on the weekends. Orwell opens *The*

Road to Wigan Pier with this statement: "Watching coal miners at work, you realize momentarily what different universes people inhabit ... The miner can stand as the type of the manual worker, not only because his work is so exaggeratedly awful, but also because it is so vitally necessary and yet so remote from our experience, so invisible, as it were, that we are capable of forgetting it as we forget the blood in our veins."

If Orwell had visited Oregon instead of northern England, my great-grandfather, Zephaniah Thomas, would have elicited a similar sentiment. With coal dust and sweat rubbed into his face, Zephaniah caught in the act of nudging mules would have stood as Orwell's icon of the miserable, his unique personality subsumed by the drudgery of his occupation. Had Orwell spent the day beside his son, Bobby, he wouldn't have noticed the long, beautiful fingers that Bobby had inherited from his mother Maud. Back then, these fingers would have loaded trams and drilled holes to blast out coal with shots. Later, those fingers would lie calmly in my grandpa's lap, the nails clean and filed, or rise elegantly to shape the words that came from his mouth so his audience would know how to interpret them. Orwell would never suppose that someday this boy would read his writings, and that Bob would write his own books, too. "On my Father's side, I am the first generation out of the coal mines. As far back as we have any record, my father's people have been miners," begins Bob in his four-page personal history contained in my binder of genealogy. Although his father, my great-grandfather Zephaniah, had little education, my great-grandmother Maud had been a schoolteacher. She had gotten a college education and was determined that her three sons would, too. Both Maud and Zephaniah died before they saw any of their sons graduate: Maud a victim of nephritis, a disease of the kidneys, and Zephaniah from a severe case of bronchial asthma developed from breathing coal dust.

Orphaned at the age of twenty-five, Bob worked as a ship welder on the docks of Portland, Oregon, pulling both the graveyard and the swing shifts so he could attend Reed College during the day. He completed his bachelor's degree at Reed and went on for a master's degree at the University of Oregon. During this time, he fell in love with his future wife, Shirley Wilkes, while they were discussing philosophy. My father Ryan was born while Bob was working on his Ph.D. at Colombia and only knew his father as an educated man. By the time Bob finished his dissertation and obtained employment at BYU, he circulated in an academic environment that was very different from the mining camps he grew up in.

This is the grandfather I remember. His house was full of bookshelves. It had a comfortable atmosphere that indicated education and a moderate income. He was an indulgent man who bought me pink soda and let me climb all over his back and head, only reprimanding me when I would pull at the remaining hair above his ears. He still worked with his hands,

Bob Thomas lecturing in Aspen Grove, UT, circa 1966

maintaining a woodshop in his basement as hobby, but we were never allowed to wear overalls at his house. To him, they were the uniform of the working class. He had crossed the chasm between blue and white collar in his lifetime so we wouldn't have to.

When I was younger, I could never understand this apparent snobbery of my grandfather. I realize now that he was proud to have been born to a coal miner, but was also proud to have moved beyond it. The lessons he learned as a boy working in the mines stayed with him. Throughout his discourses as an educator, he made many references to "the family occupation." He reflected that he "displayed both an interest in and an aptitude for digging coal. Fortunately, there were those in my life who did not insist that I choose too quickly." Later in another speech, he explained what motivated him to look beyond mining: "I was born into a family of coal miners—my father was in the mines of Carbon County by the time he was thirteen—but his vision for his sons was part of my future from earliest childhood and that vision was not bounded by what Thomases had done for generations. We were not too good to be miners, but we were

Ryan Thomas, 1979

Author sitting on her Grandpa Thomas's lap.

too good to be less than we might be. I have dug coal, and there must be some ancestral condition that makes me find the blackness of a coal mine comforting—but I yearned for a different life and worked toward it."

As I read the accounts of my ancestors and the newspaper articles from Sago, I wonder what my life would have been like if Bob Thomas had chosen differently. According to the United States Bureau of Labor, there were 78,170 Americans employed by coal mines in 2006. If Bob had passed down the tradition of coal instead of education, would my father have been the first to leave—or would I? *Leaving* was an aspiration my mining ancestors passed down from son to son, along with their knowledge of the trade. I have never viewed the occupation of coal mining otherwise. However, among these 78,170 miners across the nation, there are those who are proud of what they do, and many would not chose another profession if they could.

I realize that in approaching coal, I share more with Orwell than I do with my ancestors. As an outsider, I inspect the topic from the tunnel vision of my class and my education—one that teaches me to seek the liberation of the working classes from the hardship of their occupations. Like the miners in *The Road to Wigan Pier*, the mining families in Sago have become icons through constant media coverage. We see them as evidence of our backwardness, our persistent inhumanity, and the suffering of the working class. Yet it was their living and breathing that struck me in the news flash that first drew me in. Despite the two hundred miles between my home in Alexandria, Virginia, and theirs in Sago, West Virginia, their experiences have helped me gain a fuller and more nuanced understanding of the lives of my predecessors.

The idea of coal mining is no longer just a relic of my family's past. I cannot turn on the dishwasher, watch a movie, or write these words without thinking of coal. And, conscious of this coal, I must think of the blood in my veins, because right there from the beginning of coal—this history of our energy—my ancestors pulled it from the earth with picks and their hands. Orwell ends his introduction by asserting that "we owe the comparative decency of our lives to those who mine coal," and I owe more than most. When I think of my education, I have to remember Zephaniah, who wanted to be a doctor. He spent his youth in a mine instead of a classroom. Although he tried other occupations, he was never successful at anything else. His life working in tunnels below the earth earned my life above.

When I turn on a light in my house in Northern Virginia, in a grid powered primarily by coal, I have to consider the miners who labor in a profession my ancestors struggled in for generations and could never leave. I have to evaluate what position I have on coal, a resource on which my family subsisted and on which 78,170 American miners depend for their living. In an age when we fear the repercussions from rising global temperatures, watch the beauty of our wild lands exploited for profit, and breathe the consequences of our industry, this relationship has become complicated. In the absence of a viable alternative, coal mining is absolutely necessary to the energy demands of our society. It is a direct cost of the minutia of our everyday lives. However, despite multiple advancements in technology and safety, coal mining is an occupation in which there are still cave-ins and explosions. As long as people mine it, coal will always come at the cost of casualties.

"We all know that we must have coal, but we seldom or never remember what coal getting involves," wrote Orwell. "Here am I, sitting writing in front of my comfortable coal fire. It is only rarely, when I make a definite mental effort, that I connect this coal with the far-off labour in the mines." Like Orwell's fire, a light switch has come to represent my relationship with coal, just as my miner's lamp reminds me of my kinship with the men and women who dig it. In a way, these two light sources pivot at opposite ends of the story of coal: one represents its extraction, and the other coal's role in the electrical grid. By the illumination of a desk lamp next to my computer, I have chronicled my journey into this history of coal in an attempt to make visible what has been invisible: the lives of my mining ancestors, the lives of coal miners today, and the societal impacts of coal as an energy source. I, like other writers before me, descend to discover the workings behind a light switch. A coal miner's granddaughter, I raise a miner's lamp to reveal the intricate tunnels of a coal mine, the veins that run under the history of our modern lifestyle.

• • •

Postscript: My journey through the history of coal took place between 2006 and 2010. The events in this book are generally chronological with the narrative unfolding alongside contemporary events. The information in the chapters is current to the narrative moment. For the most recent information on relevant coal mining accidents, company mergers, and other noteworthy developments, consult the afterword.

Cymyru

2

A Welsh Coal Miner

Coal mining, in my mind, has always gone hand in hand with Welshness, because throughout my childhood I was told that I descended from *Welsh miners*—never just *miners* nor just the *Welsh*, but both integral, as if to say a particular sort of coal miner and a particular sort of Welshman. I knew it was because of our Welshness that my grandfather sang tenor in choirs (though badly) and my father quoted poetry (in contrast, beautifully). I knew that my sister Megan had a Welsh name, but nothing impressed me more about our Welshness than one Christmas when my family sat down to watch *How Green was My Valley*, and both my father and grandfather cried.

This film is based on a novel written by Richard Llewellyn, who visited his grandfather in the Valleys of South Wales as a child. Later he returned and interviewed coal miners in the Valleys, writing a total of four books based on their stories. Llewellyn so identified with these Welsh coal miners and the hills they mined that throughout his literary career he claimed to have been born in Wales himself. It was not discovered until after his death that he was born in London.

"How green was my valley then, and the valley of them that have gone," runs the last line in Llewellyn's novel. This chronicle of the Morgan family takes place in Gilfach Goch, where the main character, Huw Morgan, loses his father to the mines and his older brother to the factories. As a young girl, I did not know how closely this novel mirrored the experiences of our Welsh miners—that my grandfather must have felt a kinship with Huw, who leaves coal mining because of promising academic achievement. Huw's sister, Angarad Morgan, although in love with a preacher, marries a wealthy mine owner who makes her unhappy—Llewellyn's fictional character forming a perfect literary foil to my great-great-grandmother Margaret who, for love, spurned a rich man's son to marry a poor miner.

Margaret Davis was born in Wales in 1848 and wed my great-great-grandfather Evan Thomas in 1870. Neither wrote accounts of their lives. Margaret survives through the voices of her daughter and granddaughter, who wrote short life-sketches from their memories of her. Small scraps of detail about Evan are included in the personal histories of his grandsons and the stories of the family that have been passed on. We have no picture of Evan or even of the couple on their wedding day. Only one photograph of Margaret remains, and after searching through the plastic-covered sheets of my purple binder for the details on these two Welsh progenitors, I am left with only a vague caricature of Evan. But these female descendants preserved a fuller portrait of Margaret—that she was strong, and that she was kind. Reading these accounts, I hope there is some of her in me.

Over one hundred years after Margaret and Evan left Wales for America on the steamship *Wyoming*, I came to Wales in search of them. I watched the countryside of England move at a steady 40 mph out the window of the coach bus, each rotation of the tires on the pavement bringing me that much closer. *How green was my valley then, and the valley of them that have gone.* I could not reclaim Evan or Margaret through their own voices or their images, so I had to set foot in Wales and see the hills. I had to reclaim their earth.

Not far from the border of England and Wales we passed a towering white windmill. Modern and streamlined, the turbine looked more foreign in the country landscape than the smokestack by its side—two pillars of possibility. One sliced the air precisely; the other, stained with years of use, sent thick black smoke into the air. They rose from behind trees covering a power plant that was barely visible from the road. It seemed fitting to see a smokestack before crossing into Wales, a land whose history is intertwined with the history of steam power and coal.

The bus approached a mile-long bridge stretched over an expanse of mud scattered with puddles of seawater that rippled in the wind. Ahead a larger patch of water gleamed. It was the mouth of the River Severn that feeds into the Bristol Channel, dividing the coast of England from the coast of Wales. We neared Cardiff, the capitol and one of the major port towns developed in the eighteenth century from shipping iron and coal.

Once over the bridge, the road signs were written in two languages: *Lain Gallad*—Hard Shoulder, *Maes Awyr*—Airport, and *Gwasanaethau*—Services. This is the result of the Welsh Language Act of 1993, stipulating that English and Welsh have equal status, a move precipitated by over two thousand years of struggle for independence. In 1282 after the death of Llywelyn the Last, King Edward I of England finally subdued the Welsh kingdoms, lands known as *Cymyru* to its inhabitants, who, despite years of occupation, have maintained a national character peculiar enough to retain some aloofness. Curiously, even the name *Wales* reflects this. It means "foreigner" in Saxon.

Welsh is one of the oldest languages in Europe. There is a certain excess in the spelling; the lyric antiquity of the consonant and vowel combinations brings me back to Evan and Margaret and those who came before them. They spoke this language with an accent that has been passed down through four generations. Occasionally, I am asked if I am English or Australian by other Americans. In my verbal family, not even the New World has completely stamped out the influence of *Cymraeg*.

The bus drove through downtown Cardiff, past the castle and the university, dropping me off at the train station a fortunate twenty minutes before the next train—Merthyr Tydfil is not a popular destination, and few trains run on Sundays. In Cardiff, the villages up north are referred to as "the cities of the valley" and are looked down on for their relative lack of sophistication. In South Wales, there are five of these valleys running parallel to each other, divided by ranges of hills. Merthyr Tydfil is the largest of these towns. During the big coal years, it was the most populated industrial settlement in Wales and the setting of many important events in England's Industrial Revolution, including the introduction of the first steam locomotive.

I had come to Merthyr expecting the "old country": cottages and wooded hills where mists of Merlin lay low over trees. Instead, from the windows of the train I saw only rows of connected houses with tall chimneys clumped into the shallow valleys. A half hour later, I stepped out of the train at Merthyr Tydfil to a giant Tesco, the largest grocery chain in the United Kingdom.

The wild country of dragon lairs, sheepherders, and Druids still exists in parts of northern Wales, but the Industrial Revolution left an indelible mark on the southern region. I learned later that the connected homes are referred to as *terraced* and were built to house the workers who flocked to South Wales from western Wales, Ireland, and England to work the mines. Before the Industrial Revolution, only seven hundred people lived in Merthyr Tydfil, a small rural community that was noted for its "fairytale loveliness." By contrast, in the 1850s when the population had risen to over fifty thousand, George Borrow, an English travel writer who visited Merthyr at its industrial height, revealed the effects of 150 years of development: "all the hills around the town, some of which are very high, have a scorched and blackened look." He described the houses as "low and mean, and built of rough grey stones."

Recently, efforts have been made to improve these old miners' houses. Window frames and doorways have been painted in accent colors, and the cement walls have been overlaid with colored gravel. Potted flowers hang out front, adding a country charm to the solemn rows. Though the hills

Erin Thomas, 2006

Aberfan viewed from across the field of Merthyr Vale Colliery

are no longer black with smut, there are marks on the natural landscape that are difficult to ameliorate.

The woods, with groundcovers of brambles and wild berries, taper off at times to a similarly bumpy landscape, but in a more measly fashion. Yellow grass on top takes on an unnatural hue from the black soil that shows through. These piles of dark earth, loosely packed and eroding in places, are called *tips*, where slag from the mines has been carted out and dumped on the hillsides. And this slag, still evident on the hills, brought Merthyr County most significantly into public eye 110 years after it was the most important town in Britain during the Industrial Revolution. Like all mining towns, media documentation in Merthyr corresponds with disaster.

On my first day in Wales, I stayed with a small Welsh family with three little girls. They lived in Aberfan, a small village in Merthyr Tydfil County, five miles south of downtown. Aberfan perches on the hills across the valley from Merthyr Vale. A flat field of grass where the first shaft of the Merthyr Vale colliery was sunk in 1869 divides these two villages. Eighty years later, the mine closed and all the buildings were removed, leaving a large yellow-tinted plain of empty space where the trees stop on either side.

My first afternoon in Wales, I set out with Linda Johnson and her three girls—Emily, Naomi, and Leisle, ages three to seven—on a walk to Aberfan

Courtesy of Alan George's Old Merthyr Tydfil

Landslide in Aberfan 1966

Memorial. The day was overcast, and the tops of the hills lightly misted. Once we reached the park, which was surrounded by low stone walls, we followed a cement walk through stretches of lawn and plots of shrubbery. On seeing grass, the girls began to leap, run, and yell; their mother shushed them. She indicated an older couple who circled the park: "Girls, maybe they lost somebody." Forty years ago this used to be the location of Pantglas Junior School. Now there is a stone plaque that reads *I'r rhai a garwn, ac y galarwn o'u colli*, translated as "To those we love and miss so much." In 1997, thirty years after the disaster, Queen Elizabeth II planted a tree here; it was still a sapling. I read the marker in front of it, and Emily, the oldest, crawled behind, asking me to take a picture of her. Her long red hair trailed to her knees.

In late October 1966 at 9:15 a.m., the junior school children, many of them Emily's age, were engaged in their lessons at Pantglas, and the senior school children were arriving to start theirs. A movement started on the hill that many described as sounding like a jet plane overhead. Harold Rees, a senior school student, described it as big wave of muck higher than a house, containing boulders, trees, trams, and bricks, that "was moving fast, as fast as a car goes down town rumbling like an old train going along." On Merthyr Mountain, moisture from rain and an underground spring had saturated a tip from the Merthyr Vale shaft in the valley below. This liquefied the bottom of the coal waste, which began to move down the hill in "a dark glistening wave." In moments, this landslide of slag leveled a school wall and began to fill the classrooms, burying the children and teachers that were trying to escape. Harold watched the deluge swallow two of his classmates who were sitting on a stone wall just outside the school, plunge across the street, and knock down eighteen houses. The barber, Mr. George Williams, on his way to work that morning, was saved by a piece of corrugated sheeting. He later recalled that after the hideous rumble, there was a silence in which "you couldn't hear a bird or a child."

Within minutes, a road repair crew and miners working above the tip were at the scene. Men from the local pubs, senior school students, and mothers joined them, until there were hundreds digging through the mud and slag to save the children. Miners from the Vale Mine organized the digging, but none were recovered alive. The rescuers passed their small corpses from person to person, laying them beyond the debris and dross. The landslide killed a total of 144 people, 116 of them children.

"They say that the school collapsed first at one end," Linda says in almost a whisper, "so the children and the teachers ran to the windows at the other side. And that is how they found them, with their faces pressed against the window."

After we left the park, we took the road up to the cemetery. Leisle was tired, so I picked her up. Emily climbed and swung on rocks and railings, calling to me to come and take her picture. Naomi followed, grimacing at her shorter legs. Trees surrounded the cemetery, which rose above the city on a hill on the east side of the valley. The graves were amassed in uneven rows of monuments and tombstones. The gray light softened the edges of things and intensified the greens; the air was moist and cool. Halfway between more somber and worn-looking markers, two rows of white arches cut a straight line across the graveyard.

Under these arches were headstones that I read as I walked by—some shaped as hearts, some with statues of angels praying with their hands clasped together. "Stay off the graves," Linda called to Emily and Naomi. "People are buried under there!" The three girls had just lost their grandpa and were beginning to understand death. Leisle held my hand as we walked past the first row of arches where many of the Pantglas children were buried. Most inscriptions indicated ages between seven and ten. Biblical verses and original epitaphs were carved into many of the headstones. That of Edwin Davies Evans, age six, read: "Jesus in his bosom wears the flower that once was ours." One arch framed the stone for Robert Garfield Jones (age nine) and his grandmother Lavetie A. Jones (age sixty-one), who were buried side by side. Mrs. Jones must have been a teacher. I imagined how she must have scrambled to protect her pupils and her grandchild.

"To the memory of Dwynwen. Cherish her O Lord as we did during her short life. We miss her so." Emily swung upside down on the stair railing. Naomi cried because she had scratched her knee stumbling after her sister. Leisle's thick lower lip twitched as she whimpered, toddling by my side. A silence in which *you couldn't hear a bird or a child*—how empty Aberfan must have seemed for years afterward. Even forty years later, there must have

White arches for children killed in the Aberfan disaster.

Erin Thomas, 2006

been a reminder, a gap between those in their mid-forties and early fifties. This village had lost an entire generation. I paused at the first of the white arches and read the headstone once more: "In memory of Richard who loved light, freedom, and animals." Linda picked up Leisle, who leaned her head against her mother's chest and began to sing the theme song from *Annie*, mumbling and improvising the parts she couldn't remember.

The American photographer I. C. Rapoport arrived in Aberfan eight days after the disaster. In his New York apartment, his newborn son had lain nearby as he witnessed footage from the incident—the collapsed school, the bodies of the children, mourners dressed in the common clothing of the day. Like my experience in O'Hare Airport, a world event had taken on something personal. Rapoport arranged a contract with *Life* magazine to live in the village to document how "survivors survive." In the gray tones of his photographs, he captures the landscape of the hills, the long rows of housing, and the graveyard, but each photograph always contains a figure. In one, a woman stands with an umbrella in the distance. Flowers are strewn over a long band of newly turned earth where the double row of arches stands now. Other photographs reverberate with Welsh life in a coal mining town that had probably not changed much for 130 years—rotating around the pub, the church, and the tunnel from which the men emerged at the end of day with blackened faces.

After visiting Aberfan Memorial and the graveyard with Emily, Naomi, and Leisle, I realized that my search to understand what it meant to be a *Welsh coal miner* had much to do with these children of mining families. Though certainly one of the most devastating, Aberfan was not the first coal mining accident that involved boys and girls. Poor records were kept of the South Wales fatalities from 1788 to 1871. On one list 5,650 mining deaths are documented, but it is far from complete. South Wales was known for its treacherous mine roofs, and accidents were a weekly occurrence. Only mining incidents in which more than five men were killed on company grounds were counted, and before 1850 no systematic records were kept at all. In 1851, a mine inspector was assigned to the South Wales and Monmouthshire mines, and from this date to 1855, his tally reached 738. Boys between the ages of ten to fifteen numbered one in nine of the miners employed in these mines, but they accounted for one-fifth of the dead.

In these tunnels miles below the surface, children mined along with their parents because salaries were barely sufficient to support families. My history of Welsh coal miners is the history of one of these mining children. My great-great-grandfather, Evan Thomas, was six when he left play for the pit. In the winter months, he never saw daylight, much of his childhood submerged in the night of the mines.

3

The First Loco to Run on Rails

According to legend, a Christian princess named Tudful wandered from the northern town of Brecon to Merthyr Tydfil in the fifth century A.D. She was of the house of Prince Brychan of Brycheinoig, a half-Welsh, half-Irish king with a bounteous and educated posterity. Tudful, like a number of Brychan's thirty-six children, became known as a wandering saint for her dedication to evangelism. Now a county borough of roughly seventy square miles in the Valleys of South Wales, Merthyr Tydfil at that time was primarily populated by sheep. Here among Celtic farmers in the Taff River Valley, Tudful lived simply, founding a religious community and nurturing the sick, both man and animal. In 480, while on her way to Aberfan to visit her sister, she was killed by marauding Scottish Picts, on the rampage, apparently, since the Romans left. She fell under the sword as she knelt in prayer, the earth soaking up her martyred blood. Merthyr Tydfil owes both its name and its first mention in written history to this princess, now a Catholic saint.

Merthyr later emerges in the historical record when the English Lord of Gloucester built Morlais Castle on land claimed by the Earl of Hereford, starting a war that was settled by King Edward I in 1290. Just four years later, Morlais was seized by the Welsh rebel Madog Llywelyn, a descendant of the princes of Wales from the ancient House of Cunneda. The castle was never finished, and today all that remains are rock walls nearly overcome with grass.

The names involved in these vague histories are touched with that peculiar Welshness that evokes Arthurian oldness, and I wonder how my ancestors figured into these scattered instances of Merthyr's historical fame.

Our earliest Welsh genealogical records date to the time of Merthyr's final conquerors, not Picts or the Earl of Gloucester, but the lords of four iron companies: Dowlais, Cyfarthfa, Plymouth, and Penydarren. Our earliest knowledge of Thomases begins with the coal used to make coke for smelting iron.

Dowlais Ironworks was founded in 1759, and Cyfarthfa six years later, originally supplying bullets and canons for the British during America's Revolutionary War. Plymouth split from Cyfarthfa, and in 1784 Penydarren was founded by Anthony Bacon. Bacon soon turned over the management of the company to Samuel Homfray, whose wager was the impetus for the world's first steam locomotive.

Horse-drawn wagons initially transported iron through the Valleys to the port at Cardiff. The ironmasters of the four foundries combined interests to built a canal twenty-five miles downhill from Merthyr to Cardiff, fed by the River Taff. Barges loaded with iron floated to the ports on the canal's shallow waters. The primary shareholder was Richard Crawshay, owner of Cyfarthfa Ironworks, born of a Yorkshire farmer. When traffic on the canal increased, Crawshay did what all great industrialists would have done. He demanded preference for his boats.

Samuel Homfray of the Penydarren Works was the son of an English ironmaster. Known to have his way with the ladies, Homfray wasn't about to be beaten by a yeoman. Banding together with the other two ironmasters, he built a tramway to circumvent the canal. Ironmasters had discovered that laying tracks expedited travel. Homfray intended to transport iron by horse-drawn carriages until he had a better idea: steam.

Steam engines had been used in mining since 1712, when inventor Thomas Newcomen introduced the first practical water pump to solve the problem of drainage—the invention that began the relationship between coal mining and mechanization. Richard Trevithick, a Cornish engineer apprenticed as a boy to set up water pumps in mines, first applied steam power to vehicles. With a cousin, he developed the steam carriage, which made rounds through the streets of London. Samuel Homfray wrote Trevithick to see if he could build an engine to run on tracks. Crawshay was skeptical; Homfray swore it could be done. The two posted £500 each on the wager of transporting a shipment of iron from Penydarren to the seacoast by steam power.

In Merthyr Tydfil on February 21, 1804, in front of a crowd that likely included Evan Thomas's grandparents, Richard Trevithick showcased his "puffing steel." It was a single-barreled engine that pulled cars loaded with ten tons of iron and seventy passengers. Doubtless to the tune of hearty Welsh cheers, this first mobile steam engine crept down Tramroadside at five miles per hour, completing its ten-mile trek to Abercynon faithfully. Despite this success, Homfray abandoned the scheme. The boiler

Courtesy of *Alan George's* Old Merthyr Tydfil

Tramroadside north in Merthy Tydfil, circa 1900

broke on the return journey, and Trevithick had not addressed the issue of friction—the seven ton puffing steel broke cast iron tracks consistently from its weight and from power being applied in sharp pulses. Trevithick's steam locomotive was demoted to running a hammer in the Penydarren Ironworks. Thirty-two years later the Taff Valley Railway was constructed from Merthyr to the coast, using George Stephenson's more workable model of the steam engine and sturdier tracks, increasing both the accessibility and the demand for iron and coal. Even so, this event is still memorialized in Merthyr with this inscription: "Richard Trevithick ... Pioneer of High Pressure Steam. Built the first loco to run on rails."

By the 1830s, Merthyr Tydfil was the iron capital of the world, employing the largest number of workers and dominating trade in the production of tracks. The Welsh developed the technology for using coal to refine metal, and each of these ironworks had their own collieries to provide coke for smelting iron. Otherwise, householders mined coal for their personal cooking and heating needs.

Around the turn of the century, small collieries began to spring up for digging "sale coal." In 1824, Robert Thomas (not a relative) opened a level of coal to sell to the households in Merthyr and Cardiff. After his death, his widow Lucy took over her husband's business and began exporting coal to England, where it began to draw attention because of its nearly smokeless quality while generating large amounts of heat. For her enterprising efforts, Lucy Thomas has been dubbed the Mother of the Welsh Steam Coal Trade. Other mines were opened around Merthyr and throughout the hills of South Wales, eventually producing a total of three hundred million tons of coal from two thousand collieries.

Welsh coal was cut by three methods. In *patching*, the most ancient, coal was quarried from seams (coal deposits layered between other kinds of rock) that cropped up on the surface. Roman soldiers who first burned coal for heat would scavenge it for their fires, and Welsh farmers would later gather it for their stoves. In their collieries, the ironmasters used two methods that were more economical and less injurious to the foliage on the surface: bell pits and levels. A *bell pit* was begun by tunneling straight down to the first seam of coal. Men were lowered by a rope-and-pulley method to work on opposite walls of the pit, causing it to mushroom out on either side. The workers would send baskets of coal to the surface by the pulley system called a *wind drum*, which in some areas of the Welsh coalfields was operated exclusively by women.

Whereas a bell pit was started vertically, a *level* was cut horizontally into the side of a mountain. Miners would tunnel in and branch into corridors that led to *stalls*, or rooms, where miners would chip away at coal seams. Drawing a level was cheaper than digging down due to mining's two basic problems: ventilation and drainage. By the mid-nineteenth century, both bell pits and levels were considered primitive in Northern England's coal mines; Wales was way behind the times, but made efforts to modernize. In 1875, deep mining predominated in Wales. A colliery was initiated by sinking a shaft that became the doorway to an elaborate network of underground tunnels where hundreds of men and horses spent the majority of their lives.

Merthyr Tydfil became such a hub of industrial activity that travel writer George Borrow descended from the hills one evening in 1854 to document its activities. I imagine he used a walking stick, picking his way through the mountain brush of the South Wales hills to a valley of light and a hillside of blazes. On reaching the valley, he identified the source of the brilliance as lava-like material that zigzagged across the hill above him.

"What is all that burning stuff above my friend?" George Borrow asked a Welshman leaning against the front door of his cottage.

"Dross from the iron forges sir!"

In this cityscape that represented the throb of Welsh industry at its finest, Borrow noted some resemblance to hell. This was the home of my great-great-grandparents—Margaret Davis, who was six, and Evan Thomas, who was seven—and the people who milled about Borrow as he minced his way through the filthy streets of terraced housing and narrow alleys were their neighbors.

After navigating through downtown Merthyr Tydfil, Borrow booked lodging in one of the town's main hubs. The Castle Inn was named for Cyfartha Castle on the hills above, a nineteenth-century spectacle built by the Crawshay family, who owned the iron foundry that Borrow visited the next morning.

Courtesy of Alan George's Old Merthyr Tydfil

Cyfartha blast furnaces, 1894

The Cyfarthfa Ironworks boasted of being the largest in the whole of Britain. Of his experience there, Borrow wrote: "I saw enormous furnaces. I saw streams of molten metal. I saw a long ductile piece of red-hot iron being operated upon. I saw millions of sparks flying about. I saw an immense wheel impelled with frightful velocity by a steam engine of two hundred and forty horse power. I heard all kinds of dreadful sounds. The general effect was stunning."

At the time of Borrow's visit, the Merthyr iron trade was at its peak, and the coal trade was gaining strength in the Northern England market, Welsh coal being proven superior in trial after trial conducted by steamships. In Borrow's words, Merthyr "was till late an inconsiderable village, but is at present the greatest mining place in Britain, and may be called with much propriety the capital of iron and coal." The vast wealth pouring into this area, however, was absorbed into the fortunes of the ironmasters and collier owners. While they lived in elaborate estates, the common person's lot was meager, dark, and filthy. Merthyr by day struck Borrow as being as damned as Merthyr by night, only less magnificently hellish. He reported that it had a somewhat singed look and expounded on the satanic character of the buildings. On the people who inhabited this infernal architecture, he reflected little, stating only that they were numerous and spoke mostly Welsh. He described "throngs of savage looking people talking clamorously" and admitted that he "shrank from addressing any of them."

Merthyr, like many towns built around the iron and coal industries, attracted mostly young men at first, who, in turn, attracted the establishment of prostitution and public houses. The part of town that Borrow focused on in his descriptions was called "China" by the locals, and prostitutes, miners, and orphaned children would bunk down together in small inn rooms where occasionally the dead would go unnoticed for days.

Women would have to keep close watch on their babies, so they wouldn't be gnawed on by the rats. The residents of Merthyr were renowned for their drinking. In 1805, there were over three hundred pubs, a ratio of one to every twenty-four domestic dwellings, and in these taverns, frontier-style violence was not uncommon. Stripped to the waist, men fought duels not with guns, but in hand-to-hand combat.

Given the level of sanitation at the time, drinking alcohol was much healthier than drinking water. In 1849, Inspector T. W. Rammell conducted a public inquiry into the town, concluding: "Merthyr … sprung up rapidly from a village to a town without any precautions being taken for the removal of the increased masses of filth necessarily produced by an increased population, not even for the escape of surface water … A rural spot of considerable beauty has been transformed into a crowded and filthy manufacturing town with an amount of mortality higher than any other commercial or manufacturing town in the kingdom." According to historian Carolyn Jacob, during 1850–60, Merthyr decided to become "a proper town" and established building ordinances and water-supply regulations. Prior to this, newcomers built their houses where they pleased. High Street was a morass of mud, and sewage was pumped into a local river that would flood into the homes when it got backed up. During one month in 1849, one thousand people died of cholera.

In Borrow's account, he reports on the type of workingman one would imagine treading through the dark corridors of what George Orwell considered the most wretched job of the working class. Although factory workers performed their labors in dank factories where disease and machinery malfunction were common causes of death, only the underworld of mining threatened its workers with the dangers of cave-ins, explosions, and deadly gases. Farmhands, bent for long hours raising hoes and scythes instead of picks and driving horses along rows of crops instead of in the black halls of mines, worked in the light of day without the constant inhalation of coal dust.

The miserable living conditions of the Welsh coal towns—overcrowded housing, with piles of refuse in alleyways "beset with stinking pools and gutters"—were no worse than those of the London poor, who had open sewage lines and thick, muggy air from the soot of chimneys on side-by-side houses. The miners at least had an opportunity to rent cottages owned by the mining company that were leased at a discount to attract workers. There were advantages to mining that an outsider like Borrow would never suppose. These ranks of *savage-looking people* possessed a degree of autonomy absent from the labor of other workingmen and women.

Miners weren't paid by the hour, but by the quantity of coal they dug. Although a certain amount of weight was subtracted from their daily output on the assumption of "rubbish" being included in the tram, this gave a collier an opportunity to earn a living according to his brawn. Until the Mines Regulation Act of 1872, miners weren't tied to any particular schedule. Working days varied from eight to fourteen hours, and Welsh miners considered themselves free to come and go when they pleased. When the daily shift was officially set at twelve hours, with a half day on Saturdays, the miners exercised their liberty by taking personal holidays throughout the month—some missing up to a week. In 1868, a Dowlais trustee asserted: "The vice of the collier is to be idle, he will stay away on Monday and Tuesday and then he comes in and makes the most of the remaining four days, and he sends out as much as he can." Although it cut into their small salaries, the men bought their own picks, candles, and powder. In the stall of a mine, there was only a miner and the boy at his side. There was no shadow cast by a supervisor leaning over his shoulder—the darkness and solitude contributed to the independence of a worker and, thus, to his dignity.

Working in the mines also paid better. A woman who weighed coal at the Graig Colliery chose working at the mines over her former job as kitchen maid, claiming that although her twelve-hour shift was hard, "I prefer this work as it is not so confining and I get more money." This, of course, was during the good years. Mining was a profession particularly subject to market swings. For instance, at a low point in the market in 1859, only a regularly employed miner could depend on twenty-five to twenty-eight pence weekly, whereas after a market boom in 1862, miners could easily earn thirty-five pence a week. This was followed by a crisis four years later, where miners drew salaries below the 1859 average at twenty-four pence. This correlation between coal wages and the economy caused the famous riots of 1831 in Merthyr, where protesters were killed and an infantrymen was wounded. Protest became a constant feature of the South Wales coalfields—this tendency to unrest likely added to the Welsh collier's reputation of barbarism among the more polite upper and middle class English.

Thirty years after Borrow visited Merthyr, American writer Wirt Sikes was appointed as the US consul to Wales. In the annals of his travels, he provided a more liberal perspective on the character of these Welshmen: "The Welsh population of Merthyr is gathered in large part from the mountains and wildish valleys hereabouts, and includes some specimens of the race who (as the saying goes) have no English, with a very large number of specimens who have but little and utter it brokenly. Those of the lower class who can read … are far in advance of Englishmen of the same state in life, who often can read nothing. To hear a poor and grimy Welshman, who looks as if he might not have a thought above bread and beer, talk about the poets and poetry of his native land, ancient and modern, is an experience which, when first encountered, gives the stranger quite a shock of agreeable surprise."

Poetry was not the only thing that distinguished Welshman, but the setting of these lyrics to music. Singing was a way these miners and their families expressed community. As former agriculturalists and herders gathered from the hills into cities to work in the mines, chapels were built and formed the locus of Welsh social life outside the pubs. Here thickset and coal-stained men gathered and sang parts. Competitions among the choirs of different cities became big events, and medals were awarded to the winning choir. During strikes and, later, during the Depression, out-of-work miners went singing from door to door for donations to feed their families. At frequent town festivals, balladeers performed songs based on peasant tunes from the countryside. Merthyr was a dirty and crowded town, but a musical one.

I caught a glimpse of this one day when walking along High Street on a Sunday afternoon. I stepped into St. David's, an Anglican church built in 1847. Religious worship is on the decline in Wales, and the church was empty except for a small cluster of four men and a woman in a corner, singing different parts of a hymn. With five voices, they filled the entire structure.

My progenitors were of the grimy class of folk described so condescendingly by Borrow; they were perhaps even lower socially than those Sikes described due to their lack of any education or literacy. Of their poetic spirit, I have no indication, only that my great-great-grandfather Evan Thomas was renowned for an eloquent temper. He was the son of Frederick Thomas, who was the son of Evan Thomas. On census records the profession of each of these men is listed as "collier" or "miner." When Evan was born in 1849, his family lived on Tramroadside, the route of Trevithick's famous first steam locomotive. Evan first entered the mines at age five, clinging to his father piggyback as they were lowered down into the shaft.

Children often had to work with their fathers to support the family; a collier's wages usually did not suffice to cover the costs of living, which included rent on a miner's cottage and daily rations of bread and cheese—vegetables and bacon were Sabbath-day fare. On the Lord's day, miners were granted a bit of meat and a respite from labor. Children, most of whom were deprived of any other kind of education, could go to Sunday school.

Evan's daily labor was to load the coal his father cut from the seam into a cart, although this was a task usually given to boys much older. His hands would have been barely large enough; his coordination just developing. I can imagine his desire to feel helpful and his fear of the dark without his father close. Other children in Welsh mines would pull carts like the ones that Evan loaded, down tunnels that were so small they had to crawl, dragging the cart behind them with a harness around their waists. The youngest children, in charge of opening and closing the doors, sat in the dark and listened for the sound of horses and trams. The children were often tired, and some would fall asleep, rolling into the tracks, where they were crushed by oncoming traffic.

In 1842, having heard of the wretched conditions, two government inspectors journeyed to South Wales to interview the children. David Harris, a little boy of eight who worked in the Llancaiach mine, told the inspectors: "I have been below for two months and I don't like it. I used to go to school and I liked that best. The pit is very cold sometimes and I don't like the dark." A little girl who worked in the Plymouth mines in Merthyr Tydfil as a doorkeeper was napping against a large stone when the inspector came to speak to her. She explained that she had fallen asleep because her "lamp had gone out for want of oil. I was frightened for someone had stolen my bread and cheese. I think it was the rats." One reoccurring response among the children was that "they hadn't been hurt yet," as if this idea weighed on their minds with a sense of fear and even expectation.

After the government inspectors returned with their reports, Great Britain outlawed the employment of women and children in the mines by the Act of 1842. Depression hit the country in 1843, and many families had no way to survive without the extra income from their children. One elderly man with a large family of only daughters claimed if his three oldest were pulled from work in the mines, they would all have to be inmates of the union workhouse, a nineteenth-century institution designed to obliterate poverty. Assistance was offered, but only on condition of living and working in an establishment run like a prison. For years miners and mine owners evaded the child-labor legislation, but the practice nearly died off by the mid-fifties and sixties, when pressure came from the miners themselves to exclude boys under the age of twelve from working in the mines. Despite this historical development toward the protection of youth, it was 1855 when little Evan, either due to the poverty or ignorance of his parents, first piggybacked on his father down the mine shaft. They were lowered into the mines before the sun rose and ascended after it had set. They followed this routine seven days a week.

The lot of all Welsh children wasn't so dismal. Other boys watched flocks of sheep. Churches ran schools for poor children, and in 1870 the Education Act was passed, legislating free schooling for children between the ages of five and thirteen. William Paget, a man born in the Valleys in this year, recalls catching wild ponies in the hills, commandeering abandoned trams, swimming in the river, and going whimberry picking before he entered the mines legally at the age of twelve.

My great-great-grandmother Margaret Davis, later to become Margaret Thomas, was also a miner's daughter and lived in Pontypridd, a town, according to Welsh poet John L. Hughes, that is "nothing special": "Even the name of this place is forgettable. Pontypridd. A shamble of mystic Welshness. Pontypridd. Something to do with a bridge (there is a bridge). Pontypridd. Something to do with the earth (black stuff) … There being nothing special much around this town. Nothing at all except perhaps the river … Swilling down from Merthyr same as some kind of whip. Dirty candle-coloured by day down through Aberdare of torrents. Grunting sucking lashing whirlpools blackened through by mining trash and coal no man could burn." In this middling coal village during the age when the River Taff had just begun to take on its blackness, Margaret's mother Ann, like most coal miners' wives, must have kept her small household, pinched in a morose terraced row, scrubbed white and raw. Despite the filth of "China," many of the domestic dwellings around Merthyr were well kept. One visitor in 1869 reported seeing "tidy living rooms [with] a warm fully-furnished look. Every cottage door stands wide open and the visitor sees good coal fires, eight day clocks, sometimes, a good sofa, and a table set with glass and crockery"—in short, homes "where dirt is coined into gold." Cleaning was the bane of a collier's wife, black dust being tracked in at least once daily. I imagine there were nights that Ann cried when her husband John came home with trousers to mend and wash that were stiff and thick with sweat and coal dust. But then she would settle into a chair and callous her thumbs pushing a needle through the begrimed fabric to stitch up holes in the knees and backside. In 1858, Margaret's father died in a cave-in, most likely in a coal mine associated with the Plymouth Ironworks. I can find no record of the accident. John's passing must have been accompanied by the fatalities of fewer than four of his comrades, and thus, not been recorded—a death absorbed by a mining town's routine of loss.

Margaret, who was only ten when her father died, worked as a nanny for the wealthy households of Merthyr to help support her family. She learned to read and write while tending the children of a superintendent from one of the local mines. "Let me help you with your schoolwork," Margaret said when the children came home from school. Understanding her meaning, the children taught her to read from their schoolbooks. During this time, Margaret was baptized into the Church of Jesus Christ of Latter-day Saints (LDS church), a young and fledging religion only thirty years old in the United States. Dan Jones, a Welshman who had joined the Mormons under the leadership of Joseph Smith, Jr. in Kirtland, Ohio, and stayed with him the night before Smith's murder, had returned to his native land to convert his countrymen to this new gospel.

Dan Jones began his proselytizing in an era of religious revival. During the Industrial Age, new religions spread fast in the densely populated coalfields. Baptists, Congregationalists, Presbyterians, and Methodists were the main sects of Welsh nonconformist religion. Dan Jones's was only the most radical—instead of reformation, he claimed to preach restoration of the first century church established by Christ's apostles. Of the preachers who drew crowds to these houses of worship, Marie Trevelyan wrote: "All these great Nonconformist preachers of Wales, in person, manner and peculiarities differed from their English brethren of the pulpit, as the rugged and awe-inspiring mountains differ from smooth and undulating uplands. Their talents were many, and highly varied. Some were mightier reasoners and profound expositors—others were strong voiced thunderers, whose overwhelming appeals moved congregations to deep reflection and contrition—other possessed the pathetic power to melt and subdue, while many were gifted with brilliant imagination and vivid imagery that gave effect to apt and telling illustrations."

"Captain" Dan Jones, as they called him, enthralled audiences for three hours at a time, bringing them alternately to tears and laughter. He is reputed to have converted an entire Protestant congregation with one sermon. He led missionary efforts in Wales between the 1840s and 1850s, converting a total of 5,600. My great-great-grandmother joined later in 1861 at the age of thirteen—to the dismay of her mother and relatives, who like most Welsh, regarded the religion with suspicion.

When Margaret was sixteen, she was offered a job weighing and selling coal by Mr. Lewis, a local mine owner. This was a technical task that involved a great deal of skill and precision. Although only the poorest working women were employed in the mines, Margaret was excited by this job because it showed that Mr. Lewis trusted her. While working there, she remained active in her new religion and attended worship meetings that were held in the houses of members. At one of these cottage meetings, where religious feeling was high and Welsh voices lilted in the deep beauty of hymns, she met Evan Thomas. Four years later, they decided to marry. When Margaret informed Mr. Lewis that she was quitting her job, he reputedly told her: "My son is very fond of you. He was thinking of asking for your hand in marriage. If you will give up this silly religion and that young man you are planning to marry, I will fill your apron full with gold. My son will someday own this mine and you will never want for anything. Your life will be a success, and your home the envy of every girl in Glamorganshire."

Margaret instead married her poor LDS miner on February 21, 1870 in Saint Tydfil's Church. According to the *Merthyr Express*, the hills and rooftops were covered with snow that day and "Poor Tom the Cabby" was trampled by his horse and killed after "having passed successfully though more than one ordeal." There is no notice of the wedding, but it is evident from the contents of the paper that since George Borrow's visit, Merthyr had gentrified.

Newspaper advertisements include "Kernick's Vegetable Pills," "Artificial Teeth from H. W. Griffiths, Surgeon Dentist," and the "newest styles in Gentlemen's hats, caps, ties, collars etc. from M. Samuel." Under Local

Erin Thomas, 2008

Saint Tydfil's Church

Intelligence it mentioned that "on Wednesday evening Rock Matthews and Company gave a miscellaneous concert at the Temperance Hall, but were poorly patronized." The Mutual Improvement Class offered readings and music. These product and cultural event advertisments are evidence of a leisure class. There was even some indication that blue collar conditions had liberalized: at Bargood Coke Works, the employees were treated to a New Year's meal by their employers. Other "intelligence" revealed that the gritty aspect of Merthyr still thrived. At the Dowlais Company, five men were killed in the mine cage when the rope broke. An article entitled "Battle of the Bucket" indicated that Henry Pearce used this implement to batter his neighbor, Anna Edwards. More gruesome, the *Merthyr Express* reported that Dai Richards murdered Susannah Richards outside a public house.

Merthyr Tydfil in 1870 had plenty to offer the nouveau riche, but it is difficult to determine what actually awaited Margaret had she accepted Mr. Lewis's "golden" proposal. In the annals of the South Wales mining industry, there are many Mr. Lewises. The most likely candidate for the Mr. Lewis of Margaret's story is Elias Lewis, a man who owned several small, profitable levels in the Pentrebach area and had a son named Jenkin Lewis who was eight years older than Margaret. At that time the Elias Lewis family lived in Genthin Cottages, a property with its own grounds, which was significant, because most mining families of the day lived in terraced housing. The grounds of Genthin Cottages, however, were of modest size. Elias Lewis's visions of his up-and-coming prosperity later materialized as he upgraded to Plymouth House, a large residence with substantial grounds that belonged to one of the former ironmasters.

Before my trip to Merthyr Tydfil, I found a picture of Plymouth House online, indicating that the structure still stood, but by the time I arrived in Wales this image had vanished from the internet. At the Merthyr Central Library, Carolyn Jacob, a dedicated Welsh historian, poured over an old map, comparing it with the new. She determined the location of Plymouth House after an hour and scribbled the cross streets in my notebook.

By the time I left the library, dusk hung over the city, and I had little hope of finding Mr. Lewis's mansion. I stopped at a small convenience store across from St. Tydfil's Church, discussing my plight with the Welshmen in line. A small woman with incredibly white hair and a youthful countenance glanced at the address and calmly told me she would take me there. As it neared dark, I hurried to keep up with her rapid stride, cutting up and across a hill through neighborhoods of terraced housing from various time periods. The higher we climbed, the more I realized that I never would have found Plymouth House on my own. I asked questions, and she answered simply: unemployed, had never left Wales, lived with her parents. She left me at the stone-gated entrance to a structure that was significantly larger than the surrounding homes—too extravagant, in fact, for a single family in contemporary Wales. It had been divided into two homes, marked by differing architectural details and paint colors that met midway between two wings. "Go Grandma," I whispered, my breath freezing in the cool air.

To marry into such a family would have been a rare offer to a working woman like Margaret. Her family, headed by a widow, would have been among the poorest. Employment for women was scarce; Ann Davis might have taken in boarders or laundry, but this would have been from other miners and returned small profits. All the family's other income would have been provided by hiring out the children: the girls as household servants, the boys in the mines. I can imagine that Ann, who did not approve of Margaret's religion, put considerable pressure on her daughter to accept Mr. Lewis's proposal. Not only would it have saved Margaret from future labor, but it would also have assured her family stability and a place in society.

In the one portrait we have of Margaret, she appears to be in her forties. Her hair is pulled back tightly from a square face with broadly cut features. Even lifting the corners of her mouth that had begun to sag, erasing the lines under her eyes, or rounding her cheeks would not make her a striking woman. At a glance there is nothing that distinguishes her from the myriad black and white images that illustrate the past, but there

Erin Thomas, 2008

Plymouth House

is a softness in her eyes that suffuses her whole expression. The only physical description that survives in our records is that these eyes were intensely blue. There must have been something behind this ordinary face that was remarkable: unique enough to draw the admiration of a rich young man, and courageous enough to turn him down for a religion and a poor miner that she loved better.

Margaret's daughter Mary Jane wrote the history of her mother, and it is through her that the story of Mr. Lewis has been passed down. In a life sketch of no more than a page, Mary Jane has included this episode in detail. Margaret must have repeated it to her children often. Margaret's family always struggled financially; perhaps she retold it to remind herself and her children that love was more valuable than comfort.

After marrying in the Saint Tydfil's Church, Margaret and Evan moved to Pentrebach, where Evan continued to work in the mines. Margaret taught Evan to read from a Book of Mormon given to them by missionaries. They raised five children in Wales, and in 1874 they sailed with Evan's father Frederick to join the growing LDS community in Utah. *Pa bryd y cawn fyned i Seion?*—"When may we go to Zion?"—many of the Welsh would ask the missionaries who baptized them. By the end of the nineteenth century, twelve thousand Welsh had converted to this new religion, the majority

Margaret Thomas, circa 1900-1910

from Merthyr Tydfil, and five thousand had immigrated to America, many forming the core of the now world famous Mormon Tabernacle Choir. This Welsh influx represents an estimated 20 percent of the gene pool of Utah, according to the website "Welsh Mormon History".

I can only imagine the anticipation of Margaret and Evan as they hurried their five children up the plank to board the *SS Wyoming*. They would have had few belongings to carry with them to the New World: the clothes on their backs, a dish or two from their terraced house in Merthyr, and leather-bound scriptures wrapped in a rag. After leaving the fiery furnaces and charred hills of Merthyr, they hoped to find a better life with the Saints. Evan, on leaving Wales, meant never to descend down that mine shaft again.

4

Two Miners' Sons

Although Emylyn Davis was short and wide, he sat thinly, a passive presence in the middle of a sofa in a small, tidy living room with lace curtains and broken-in furniture. His wife, Edna Davis, an entirely practical and talkative person, rested her elbows on the arm of a flowered armchair that contained her neatly, down to the crease in her blue polyester slacks.

They seemed to be somewhere in their seventies; their company and the comfortable tatter of their living room reminded me of visiting my grandparents. In the local LDS congregation that the Davises attended, Edna was known for her genealogical efforts, a highly valued skill in the Mormon church, where a primary focus is placed on our ancestors. Although my family has departed from the tradition of coal established by our Welsh predecessors, we have held to the faith adopted by Margaret and Evan. The Robinsons, a Mormon couple I stayed with for a portion of my trip to Wales, recommended I speak with the Davises because there was some possibility that Emylyn might be related to my great-great-grandmother Margaret.

In Wales, to be a Davis or a Davies is as common as being a Thomas or a Morgan. Children used to take their father's first name for a surname, which—compounded by the geographic and cultural isolation of Wales for so many centuries—resulted in very little nominal diversity. Though it was possible Emylyn and I were of the same Davis, it was equally possible that we were not. There was no indication that any of Margaret's immediate family ever joined the LDS church, and Emylyn came from a long line of active Latter Day Saints.

I knew nothing of the Davises left behind after Margaret emigrated with her husband Evan to America. On what terms, I wondered, did she leave her family? She must have been a sore disappointment to her mother Ann, as the daughter who spurned the family's one opportunity at prosperity.

Was it possible that any of Margaret's siblings eventually became reconciled to the religion that ruined the Davises' financial prospects?

I threw out names and dates. Edna shook her head; she knew of no connections. Preparing to take my leave, I told the Davises the purpose of my research, "I'm writing a book about coal mining."

"Oh, Emylyn's father was a coal miner." Edna looked across at her husband, who had stayed silent for most of the discussion on his pedigree. A stout man in deteriorating health, Emylyn hardly twitched, but the mention of coal mining animated his face. If this man was my relative, his father would have been Margaret's nephew. Her brothers probably never ventured far from the family occupation. His father would have lived the life Evan and Margaret left behind.

At this cue from his wife, Emylyn opened his pale, dry lips and began to spin a monologue: "My father worked a seam of twenty-four inches that ran for miles." In a calm, even cadence and a clear Welsh accent, Emylyn told me about the mines—stories his father must have recited to him that had now been tailored by his own telling. He was a good and steady raconteur, having obviously practiced on his own children and grandchildren: "Boys in their mid-teens would work seams as thin as eighteen inches. My father would lie on his side with a pickax and a mandrill, pulling the coal out with his hands, breathing in coal dust seven to eight hours a day. His only light came from an open flame lantern."

In 1881, a miner from Rhymni, chronicled in a volume assembled by the National Museum of Wales, explained how managers would respond to those who complained about the conditions under which Emylyn's father mined coal: "You chew coal. It does you good. It cures healban. It'll cure chest troubles."

As early as 1830, coat dust was known to cause black lung, or, as it was called in the nineteenth century, *miner's asthma*. It is unlikely that the manager the miner quoted above was unaware of this at that time. Although numerous miners died from cave-ins, according to one estimate this accounted for only 17 percent of all deaths. Others died from mining-related diseases, among them black lung and kidney disease from the rats.

"There was no sanitation," Emylyn continued. "Not to be rough," he nodded in my direction several times, "but the miners would have to relieve themselves in the mines, where the rats scuttled around on the ground. When the miners would cut themselves, because there was no way of cleaning the wounds, they would urinate on themselves. This healed things nicely."

"In those days, they lowered the men into the mine in cages. They called it entering 'the bowels of the earth,'" explained Emylyn. Sometimes the men would be crushed by the machinery. Once below, there were other dangers. The two methods of deep mining in Emylyn's father's generation would have been room and pillar and longwall, both of which are still used currently. In the room and pillar method, rooms were mined, leaving pillars of coal in between to support the roof. A lot of coal was lost in these pillars. After a whole area was mined out, miners would *retreat mine*, taking out the pillars and leaving the roof to collapse behind them. Longwall mining was less common in Wales; it involved working a whole coal face, propping up the roof behind with piles of rubbish as progress was made forward on the seam.

Wood and metal props were used to brace roofs every morning before mining, but cave-ins were a daily occurrence. Other hazards included flooding and the more insidious danger from gases: choke damp, white damp, fire damp (carbon dioxide, carbon monoxide, and methane), as the Welsh miners classified them, each according to their menace. White damp—odorless, tasteless, and colorless—was the most feared.

"The men would bring canaries into the mines, and when the canaries fell from their roosts, they would pull each other out, but they would leave the horses. When the air was clear, they would clean out the dead, cutting the horses into pieces to send them up out of the mine."

"Sometimes the men would be afraid. They would sing Welsh hymns, 'Nearer My God to Thee'—ten to twelve in harmony." Emylyn focused his eyes in the distance when he said this.

I was relieved to know that fear was something the miners could admit to each other—one man striking out into the dark with a note. Although they couldn't see each other, their voices joined together and echoed through the dark tunnels. There between carbon walls made of creatures a million years dead, these voices signaled that there were living men. My father sings with the deep, rich voice of a Welshman; there is strength in this sort of music.

"My father came home all black and would wash in a wooden tub," Emylyn told me. These tubs were in the middle of living rooms and made from beer barrels sawed in half. "A miner would strip to the half and wash the top part of his body. Then he would cover and strip to the other half, washing the lower part of his body"—preserving a semblance of modesty in a family thoroughfare. "Don't wash my back," Emylyn's father would say to his wife. "The men believed it would weaken them," Emylyn explained.

I imagined Emylyn's father and the Davises before them relying on the strength of their backs for the meager living it provided for their families. This must have been quite a weight pressing down on a man heading into the mines, pickax in hand, always toward the possibility of death.

"You could always tell a miner because they had black marks on their face[s]," Emylyn said, touching his own lightly. The scratches would turn blue because the coal dust became imbedded in their skin, the coal never quite scrubbing off. A coal miner was marked for life.

"When we had a son, Alan, we decided he would never go underneath," Edna commented from the easy chair beside me.

Emylyn nodded. Edna suddenly straightened her back, seeing something through the white lace of her living room curtains. "There's Malcolm going by," she said to Emylyn and then stood up, walking swiftly across the room down the short entranceway to the front door.

"Hullo there, Malcolm, we've got a girl here from America!" I heard Edna call from outside. A fit man, looking to be in his sixties, entered the room and sat gamely on the couch.

Returning to her armchair, Edna explained, "Malcolm was a miner." When the Davises began to exchange greetings with Malcolm, I realized that they had been using "foreigner" English with me. The variety tossed quite loosely and joyfully between them seemed to be a compromise between the Queen's English and *Cymraeg.* I could make out almost half of what Edna said, but Emylyn and Malcolm's communication reached my ears as a jumble of cheerful gibberish.

Pausing, Edna looked over at me as if she had just remembered, consumed momentarily by Welsh pleasantries, "If you don't understand us, just say." And Malcolm slapped his hands on his knees, and said, "So?" to indicate he was ready to answer questions.

Malcolm was perhaps the most robust-looking elderly man I had ever seen. His face was remarkably unlined for his age, and certainly not in blue. There were no visible signs of a life lived for a good portion underground. Malcolm sat erect, and I wondered if perhaps a black back was the secret to his posture, years of dust and sweat rubbed into a polish, fortifying and holding up the old sinews.

"You were a coal miner?"

"Yes, worked in South Wales, rode the train 1,800 feet in; we worked six miles underground."

Malcolm, unlike Emylyn, was straightforward—his story weaving lacked the renowned Welsh poetic streak.

"I was a coal-face fitter, a mechanic."

Like my Welsh predecessors, Malcolm was from a line of coal miners, but his father insisted that his sons not follow him: "My father came out in 1947. 'I'm the last one to go down that,' he said. I became a miner, my son did too."

"Your son is a miner?"

"Yes, been there since he was seventeen, left school."

"He works in the Tower Colliery," piped up Edna.

The Tower Colliery is located near Merthyr and has a compelling history. In the early 1980s, Margaret Thatcher began to make threats of closing British minessince cheaper coal could be purchased from Spain. The National Union of Miners called for a strike in 1984, when the shutdown of twenty mines was announced. After a year of political wrangling and violence, the strike ended and mines began to close rapidly, putting two hundred thousand men out of work. Although there were varying levels of involvement in other parts of the United Kingdom, 96 percent of the South Wales miners went on strike, and 93 percent held out until the end. Throughout the conflict, Thatcher characterized the striking miners as "the enemy within."

The Tower Colliery was one of the last mines to be closed, ceasing to operate in 1994. A year later, the miners pooled their severance pay, buying the mine from the government for £2,000,000. Although a small victory for an industry that was all but obsolete, I appreciated the symbolic heroism of their act. For most of Welsh mining history, miners had been at the mercy of powerful ironmasters and collier owners. After 1947, when the mines were nationalized, they were at the mercy of a government that antagonized them for fighting to keep their jobs. In the Tower Colliery, each miner owned a portion of the mine, and the money from the coal each miner dug went directly to his own pocket.

"It's got two years left, and then they'll have to close it up." Malcolm reported cheerfully.

"What will your son do?"

"Don't know; he is forty-seven. Don't you want to know how much I made?" Malcolm burst out, as if this was the question he'd been longing to answer the whole time.

"Sure."

"In 1958, I had two children. I made nine pounds a week. In 1987, I made fifty pounds a week." According to the current exchange rate, this equals approximately eighteen dollars and ninety-nine dollars, respectively.

He said this with a rush of blood to his cheeks. Proud of this fact, he sat up more erectly on the couch. I couldn't tell if it was due to the increase in his pay or the meagerness with which he was able to support a family.

"We would get eight loads of coal a year as concession. They brought it in front of your house, right outside. We'd save the residue—bedlam coal—all the little bits. Sell it. And each year we'd go to the seaside."

There was no bitterness in Malcolm's evaluation of his life, a life my predecessors did not choose and were always looking for a way out of. Although his father had different plans for him, Malcolm went down the mine shaft. Perhaps for similar reasons to the one a thirteen-year-old boy from Garndiffaith expressed in 1908: "I had plenty of opportunities to stay on in school, but all I wanted to do was go in the pits because my buddies were there." Or, like my grandfather Bob Thomas, perhaps Malcolm had shown an aptitude and interest for coal mining in his youth, but unlike my grandfather had decided it would suit him better than another career.

"There is sadness everywhere," Malcolm remarked, as if he could tell what was on my mind.

"Yes," I nodded back to him. Mining was dangerous, but so was living. Malcolm's cheerfulness was evidence that there was also satisfaction in his choice—each life path offering its occasional stipends of bedlam. I turned this idea over in my head while Malcolm slapped his knees and exited as abruptly as he had entered. Emylyn sent farewells after him in the low, musical tones of a hollow reed. Two miners' sons had told me different stories of coal. Despite the disparity in their health, likely they were no more than five years apart. From his account, Malcolm couldn't have been any younger than seventy. One had left the profession; the other had followed his father and had been followed by his son in turn. One related the story of accidents, sicknesses, and burdens, the story of coal I was familiar with—the story of the Davises and Thomases as it had been handed down to me. The other, like an honest workman, had reported the details of his trade. One filled the room with the loveliness of his voice, and the other with his marvelous vigor.

5

The Paths of Blind Horses

After days of cloud cover and brooding, Merthyr Tydfil opened its eyes, blinking at the warm rays of the morning. I stepped out to the road, walking past the rows of houses with brick, stone, and gravel facades. In the backyards, laundry lines and odd pieces of junk were just visible above the fences when viewed from the high road near the mountain. The houses tilted toward the center of the slender valley, where even the dry field over the old colliery looked inviting. I was on my way to take the train to Merthyr Tydfil's town center to do some research in the library, but I heard the rumble of the departing train when I was still a block from the tracks.

Returning to wait for the bus, I glanced at the hill that sloped up from the other side of road. Grass was growing where houses once stood. Since the mines closed, there had been a lull in real estate. A group of rather roguish looking sheep grazed on a tilted patch across the street from me, looking as if they had been sheared with a chainsaw. Before the energy possibilities of Merthyr were discovered and the Dowlais Company was founded in 1759, a mere forty families lived here, planting crops and raising animals. I couldn't help but think that before the coal and drudgery of Evan and Margaret, one of my ancestors climbed up these hills on a morning like this. Merthyr Tydfil had been conquered by the Picts, the English, and then the ironmasters; it was my turn to climb a hill and stake the claim of my family's inheritance.

In my knee-length skirt, I heeded the call of the hills, which had been beckoning to me since I arrived in Wales. I scrambled over the low rock wall that held in the hillside and went up through the mountain brambles, scratching my legs and wetting them with dew. I gazed across the clearing in the middle of the valley to Aberfan and the graveyard with its rows of white arches. The sun warmed my head and made the dew on my legs sparkle. The sound of whistling drifted over the tangled growth, and I imagined it was the shepherd of the homely little flock below, restored to his rural calling of nearly 250 years ago, looking over the pleasant valley no longer stained with smut and fire.

Leaping and sliding down the hillside, I watched for the bus through the trees, keeping my ears pricked for the sound of tires on the pavement. Once through the trees, I was faced with Malcolm. He and a few elderly ladies waiting at the bus stop stared as I tried to negotiate the stone wall gracefully.

"We saw you up the side of the mountain," Malcolm began when I reached them.

"Yes."

"In these parts, we are not used, to seeing women, up the mountain, in the morning." He paused between each cluster of words, his voice lilting up at the end as if he was reciting verse. A deep and thoughtful look passed over his eyes, as if he had stumbled upon a profundity. If Malcolm had not entirely caught the poetic sentiment of his countrymen, he had certainly inherited the rhythm.

The nervous twitching and grimaces of the elderly ladies made me realize that my attempt at bonding with the hills of Wales had constituted a cultural faux pas, or at least an abnormality. "Yes, I am a mountain girl," I said stupidly, realizing how close this was to *mountain goat* after it had left my mouth. I didn't know how to explain my need to embrace the hilly land that my ancestors had left and the desire to meld it with the rugged land they came to. Margaret and Evan arrived in Carbon County, Utah, in 1886. Later, after living in Oregon and New York City, Bob Thomas moved back to Provo, Utah, in 1951. Here my father was raised and met my mother. My childhood was spent wandering the mountain face above my maternal grandparents' house on the foothills, which was less than a mile from the home of Bob and his wife Shirley. My parents settled farther north in Utah Valley, and during my teenage years I would ride my bike up the canyon trail in the summers, rock climb without ropes, and scale the Bridal Veil waterfall. Knowing and belonging for me has as much to do with place as it does people. My spontaneous morning scramble was an important step on my journey toward Margaret and Evan. The Rockies in Utah and the Valleys in Wales—these were family terrain and my only truly physical link to my ancestors. I didn't know how to explain this to Malcolm or the elderly ladies who eyed me askance, so I turned and waited for the bus.

Once in town, I stepped out of the bus almost directly into an open-air mall. I browsed through the shops, noticing how I contrasted with the

Erin Thomas, 2006

Downtown Merthyr Tydfil

short, dark, full-figured women and thick men. I can claim Wales as the land of my primary ethnicity, but tall, thin, and fair, I drew mostly Swedish traits out of the family gene pool. Compared with the typical Welsh wide jaws and square faces, the lines of my face are angularly Italian.

"Look how tall she is," I heard an elderly woman cackle behind me.

I was seeking reunion with Wales, with Evan and Margaret, but I felt like George Borrow: after having descended into a foreign land, I found myself somewhat at odds with the local citizenry. A market blocked the road in one of the main drags, with most venders selling cheap household implements and poorly made clothing. I had chosen to wear a corduroy skirt to avoid the look of an American tourist while still appearing casual. Jeans and a T-shirt would have blended in much better. I stopped in a retail outlet and bought a pair of socks with *Cymyru* written on the soles.

High Street in Merthyr Tydfil,
WHS Kingsway, 1910

The buildings downtown, like the miners' houses, had fresh coats of paint. Flowers hung from lampposts, and the sun moved in and out from white clouds—a very different image than a photograph from 1910 I had found on a Welsh genealogical site on the Internet. In this photograph, the atmospheric effect on the buildings in the distance is significant. A stormy day could have caused this, but the people are hardly blurred, indicating a quick snap of the shutter. It must have been bright, with the clouding of the buildings resulting from the particulate matter pumped into the air by the smokestacks. Puddles gather on the street, and the tracks of a streetcar narrow into the distance. The photograph is of High Street, and men and women dressed in black line the walks next to the storefronts. In Evan and Margaret's time, these streets would have been even more densely populated. Between the hours of six and ten p.m., the working class would emerge in hordes for their evening promenade: girls in canvas dresses covering woolen petticoats, colliers with tin boxes and broad-brimmed hats. Women in checked shawls would sell their eggs, butter, and bacon. Apples and herring were sold in barrels. Amid the food retailers, quacks would call out their newest remedies for ailments ranging from indigestion to immoderate passions. Ballad singers would dispense a tune at the price of one pence per song.

I turned onto High Street from the retail outlet stores toward the library, past St. David's Church. There was none of the nineteenth-century plebian clamor; the street seemed deserted, hardly a thoroughfare. Only a group of teenagers slouched in front of the central library, where a statue of Henry Seymour Berry, Baron Buckland of Bullich, stood. A

Courtesy of Alan George's Old Merthyr Tydfil

Cyfartha Castle

rich collier owner, he was granted a peerage in 1926, but inconveniently fell from his horse and died two years later, never fathering a son, thus beginning and ending the Lords of Berry. I walked past this coal-made man to the library's collections, pursuing the history of the nameless.

Hours later I drove with John and Sandie Robinson to Big Pit, on the edge of the town of Blaenavon, twenty minutes from the city center. A shaft was sunk here in 1860, and Big Pit operated until 1985, when the majority of the mines across Wales were closed by Margaret Thatcher. In 2001 the government offered funds to convert Big Pit into a museum, and some of the men who had previously dug its coal were now employed as tour guides. This was consolation to only a handful of the thousands who were impacted by Thatcher's decision. Since the nationwide closing of coal mines, the population of Merthyr Tydfil County has declined steadily, with several hundred residents leaving each year.

Big Pit was an hour out of Merthyr, but the drive itself was a tour in South Wales' coal history. Just outside of Merthyr town center, to the east of the road, loomed an enormous viaduct of seven arches, 455 feet long and 92 feet high. The Pontsarn Viaduct was built in the eighteenth century in order to carry water to the canals—a method initially used to transport coal and iron. When the railroad was constructed, tracks were laid right over its top. Cyfartha Castle resided on the mountain face above the viaduct. Built by William Crawshay II in 1824, the style of architecture was hundreds of years out of date at the time. The obvious reference to the feudal period was intended. Crawshay, an ironmaster of the Cyfartha works, was extravagant in his tastes, but generally fair to his workers.

Erin Thomas, 2006

Big Pit Mining Museum

"It is a monument to a man's pride," Sandie explained. "The ironmaster was king. The girls knew it. If he wanted them, he had them."

William II was the grandson of Richard Crawshay, a farmer's son who rode his pony to London at the age of fifteen after an argument with his father. Selling the pony for an apprenticeship in an iron warehouse, Richard rose through the ranks to establish the world's most successful iron dynasty, handing it down through three generations of sons. William II paid £30,000 for Cyfartha Castle and lived there through three wives and the fathering of fifteen children. There are no records that justify Sandie's accusation of philandering, but her comment is evidence of the power the ironmasters and coal owners wielded over their workers. I remembered Margaret and was struck once more by her courage in refusing Mr. Lewis's offer.

At Big Pit, I waited on a long wooden bench for the next tour. An English family waited with me; their two young girls, with wind-tossed hair down to their shoulders, turned their knees in and fidgeted quietly. A couple in their midthirties smiled at each other and us, but said little. In assembly-line style, we took off our watches and surrendered our purses and backpacks to put on hardhats, belts, and heavy oxygen flasks. The last thing we strapped onto our hats was a miner's lamp. Our tour guide, Huey, a congenial man with a twisted nose and a receding knob of a chin, cracked jokes. Dressed in an orange jumpsuit, he was the genuine article: a retired miner. We went down the shaft in an elevator that seemed to drop fast but was moving relatively slowly.

Stepping from the elevator to the mine entrance, we stood for a minute in what the Welsh miners would call a *slae weld*, or a break for seeing. In the summer, when the sun rose before the men went to work, they put their heads down and talked to one another until they recognized the people who were passing, a signal that their eyes had become accustomed to the dark.

Once my eyes adjusted, I sensed a cave-like coolness. I could hear water dripping, and the light from our lamps reflected against slick walls. I felt what the miners meant by *the bowels of the earth*. In a sense, walking in a coal mine was like being swallowed. The stone seeped, suggesting the digestive juices of a black stomach. I understood why, on every descent, miners considered the possibility of never returning to sunlight.

Huey indicated a large metal cart full of coal. "They used to pull the trams along the tracks here, and each weighed a ton. What do you think is heavier? A ton of coal or a ton of iron?"

The girls giggled, but neither offered an answer.

Huey bent down so his head was on their level. "Alright then, right or left?"

The younger girl pointed and we went right, down a tunnel supported by boards. I had to crouch or the top of my hardhat would rattle against the roof. When we got past the first door, into a hallway between two doors, our guide instructed us to turn off our lamps. He told us about the six- to nine-year-old children who used to watch the doors in the dark. At Big Pit the older children, aged ten to twelve, pushed trams, and the boys, who were twelve to fourteen, helped their fathers fill six trams a day. I held my hand in front of my eyes, but I couldn't even see the outline of my fingers. I thought of Evan, a small boy in such a black world, and sudden tears gathered in the corner of my eyes. He had never resigned himself to mining. He must have resented the poverty that pulled him from school and the hills to this darkness and the fears of scuttling rats and rumbles of shifting rock.

"Alright, turn 'em back on now."

I wiped the tears from the corners of my eyes, and we continued through the tunnel until Huey stopped us. We focused our lamps on a room off to the side, where there was a seam of coal little more than a foot wide. The seam had been carved out of the mountainside, but the wall above it was still intact.

"Miners would lay on their sides 'ere, with a pick and chip out the coal."

Manual labor was something I understood. I had worked for the US Forest Service for two summers after I graduated from high school. I built fences, using axes, shovels, trowels, hammers, and drills for ten hours a day. I wore gloves and a hardhat, carried fenceposts on my shoulders up hills, and hauled tree trunks through thick forest. Occasionally we cut trail, swinging axes and trowels over our shoulders in a rhythmic action for hours without breaks. But my work environment was mountain air and sun; wet days only emphasized the breathing green beauty of a forest. I enjoyed it, the sweat and the out-of-doors, but I knew that work like this made people old fast. After my two summers, I didn't apply again.

Looking down at the twelve-inch seam, I thought of the movement and muscle that would be used to direct an implement such as a mandrill. I could imagine swinging it on my side; the strokes would be cramped and stilted, the coal much less yielding than the earth I had turned with the trowel and the roots I had cut with my ax. Muscles would ache intensely from the same posture day in and day out. Inhaling coal dust and being cramped between dripping mine walls where rats scurried up and down would sap out any pleasure I felt from my strength and hard work. I realized this was one thing I did not inherit from my Grandpa Thomas; the darkness of a mine gave me no comfort.

We moved from the seam to a section of wooden stalls built in a room carved out of the tunnel.

"This is where we kept the horses. We still have one," Huey chuckled.

We walked past a plastic horse with a silly grin. Smaller than life-sized, it made the stall it stood in loom larger. Abbot, Bounce, Victor—the names of the horses were written on the stalls, and I wondered about these animals that lived the better part of their lives underground.

"The miners loved the horses, would feed them food from their lunches, and brandy from their flasks. Some even pulled the flasks outa miner's pockets with their teeth."

Huey cleared his throat. "The horses were brought down when they were four," he said as he put his hands in the pockets of his orange jumpsuit, leaned forward slightly, and rocked back, "and worked for ten years."

Around the turn of the century, seventeen thousand horses worked in the South Wales coalfields, but it was more difficult to get a horse down a mine shaft than a man. A horse's legs had to be roped to its body, and men would lift it into the elevator. Due to this challenge, in some mines horses were only brought out for a couple of weeks a year, and during explosions, they were left behind. When the horses were brought up into daylight, they would run around the pastures and kick like mad. In some mines the horses were never brought to the surface at all. These horses lived out their lives in darkness, and when they finally were let out into the light, they were blinded by the brightness.

"What happened to them afterward?" I blurted.

"I dun know," the guide admitted. "By the time I started working in the sixties, we no longer used horses."

Throughout the tour, we walked no more than a mile or two in a mine that covered a total of thirteen square miles. Nearing the elevator to the

Erin Thomas, 2006

Miner's terraced housing in Merthyr Vale

world above, Huey passed his fellow guides who were calling back and forth in French, German, and Russian, mocking the expectations of the tourists who judged them as insular. They must have taken people from all over the world through these passages of their former occupation.

We passed a short, jolly man leading his group into the tunnels.

"'Ello, you know Huey, if I could be reborn as anything, it would be a pigeon."

Huey chuckled and kept moving.

"Want to know why?" the other guide called to ours.

"I know why," Huey responded, still laughing to himself at his comrade who had worked all his life underground and wished to defecate from the sky.

On the way home, the Robinsons dropped me off at their friend Anna's home. I had been told all day that she and her husband "live in an old miner's house." Flat-fronted rows stretched out on either side of a narrow road with cars parallel parked on one side, so there was barely room to drive a vehicle along it. The fronts of the houses on one side faced the back doors of the houses across the street. Anna opened the door with her son Malachi on her hip; her husband Nathan stepped up beside her. They made a handsome couple, standing in the small opening to their home: she pretty and blonde, with an open smile, and her husband stocky, with a square face. True friendliness shone from both of their faces. Anna told me it had been a "manic" day, but everything had settled down all right. Her eldest son Jacob hurried from behind his mother's legs, ran into the house, and buried his head in the couch.

"He's shy," Anna explained.

They directed me toward a table, and Anna brought out a noodle and cheese casserole. Jacob refused to come and eat, and Malachi began to holler and growl aimlessly. Anna tried to calm him while Nathan sat down to entertain me.

"So this is an old miner's house?"

"Yes, we added this back part on. There was only one small room on top of another. It used to house a family of ten."

While the commotion of two small children filled the living room space, I tried to imagine the atmosphere of Evan and Margaret's home. They had had five children when they lived in Wales. Anna joined us, and Nathan told me how they had met at a regional dance for all the young LDS singles in the northern part of the country. Anna was from England, and Nathan was a native of Wales, where he owned a small landscaping business. They had been married four years ago, both at twenty-one, nearly the same age Evan and Margaret were when they began their lives together.

"When I met him, I didn't understand a word," Anna teased.

Nathan blushed. Anna looked happy in the life she'd married into: a small miner's house, and two sons—calm Jacob and vocal Malachi, who had continued to test his lungs throughout the conversation. Margaret had made her home in a place similar to this one instead of at Plymouth House, the grand residence that looked down on the terraced housing from above. Evan must have been a Welshman like Nathan, bright-countenanced and eager. Most days he must have been gentle, with only injustice inciting his tongue and making the tips of his ears turn red. There must have been some happiness lived in these small houses.

Anna stood up to settle a disagreement with the boys, and Nathan cocked his head and said abruptly: "Let me tell you one thing. My father told me this. When the horses had worked ten years in the mine, they would retire them. They'd bring them up from the mine and release them into a field at night. When it was daylight they would be blinded, so they would walk the floor plan of the mine, wearing a path into the grass by walking it over and over."

It was the answer to the question that had been disturbing my mind since my visit to Big Pit. I imagined the paths in the pasture marking the tunnels that lay below with the horses trudging at a measured pace, as they had every day for the last ten years of their lives.

"Because they were blind," I repeated.

"Yes," Nathan nodded. "Because it was all they knew."

There was something about the silent suffering and the burdensome routine of a blind animal that was emblematic of the men, women, and children who labored with them. Helplessness is epitomized in the image of a colt with its legs tied to its body, hefted into the cage by a man on either side, its head tossing back and forth, nostrils snorting in fear and dismay as it is lowered down. I wondered for how long a horse remembered the fields and sky before its vision was dulled by the dark passages and brief shadows cast by headlamps. By the time the animal was roped and carried aboveground, fearful and confused once again, it only had a memory of darkness and the tunnels of a mine.

I think of the path beaten into a field by Evan's father Frederick, his sons, and the grandsons who trailed after him. In the words of an Onllwyn miner born in 1887, "There was nothing else here. Nothing. What else did you have? It was only the coal mine." Evan and Zephaniah followed their fathers into the mines, not because of blindness—they longed for a better life—but because it was all they knew. From the circumstances of their early lives, the only option that lay before them was a trail leading down the shaft of a coal mine. "A Miner's Catechism," written in 1844 by an anonymous Welshmen, reflects this reality:

> Q. What is your name?
> Peter Poverty
>
> Q. Who gave you that name?
> My godfathers and my godmothers in my baptism, wherein I was made a member of the Black Coal Pit, a child of poverty and an inheritor of the sunless mine.

For my grandfathers, it was always the dilemma of a son who could not go to school. One after another, each in turn, was obligated to supplement the earnings of his collier father, a march of poverty tread by Thomases for generations, until Robert and his brothers stepped out.

Carbon County

6

The Rattle of Dead Men's Skin

I grew up under the shadow of a great mountain. The slopes that rose above my cul-de-sac in Orem, Utah, conjure up deep familial love, but also reverence. No matter how intimately you get to know the pink light that kisses the snow on the peaks during a winter sunset, or the spill of radiant light over the dark contours of the morning, the Rockies always keep a distance. They are too grand and too wild to be household mountains. They do not belong to the city that climbs into their foothills below; the city belongs to them. Farther south, the black limestone becomes tawny colored sandstone, and the rugged face of the mountain lands is not chipped by ice and ground by glacier, but carved by the wind and the rain.

There is a sense of antiquity about the arid mountains of the American West that is unlike the slender valleys and old hills of Wales. In central Utah, the prehistoric Fremont kept their living sparse, and the silence of the pictographs they left on rock walls merges with the silence of petrified dinosaur bone. Shale is scattered with fossils of sea things that speak in codes we are just beginning to crack. The soft bodies of these creatures sank to the bottom of Lake Bonneville, combining with plant refuse that, after years of compression, became coal. Miners who dig out the seams sandwiched between the sediment of these mountain ranges occasionally turn up a Brontosaurus footprint.

Carbon County contains the relics of many beginnings, including my own. I remember crawling over a big pile of shale during a summer camp my school district sponsored for fifth-grade students. We were in Scofield, near an opening to an old mine level, the location of the mining accident of 1900 the teachers had elaborated on during a hike that seemed too hot and long. I had already been told this story by my parents and was looking forward to poking my nose into a dark tunnel, but the opening had been barricaded with stones and boards. Not even a sign had been erected by the park service or the historical society to mark the famous disaster, but I seized the opportunity anyway. "My ancestors died in that mine," I bragged to a girl and a couple of boys that were sorting through the shale pile beside me for fossilized trilobites.

Before the hike, we had visited a mining museum, hardly more than a cabin that displayed various objects retrieved from the explosion. All I remember was a boot. The curator cheerfully explained that the owner had died in the accident and left sloughed-off pieces of his skin inside. My mind seized on this one morbid detail. Ever since, I have not remembered this experience without imagining the rattle of dried skin in an old leather boot.

Consequently, my family's history in central Utah has never been separate from the idea of disaster and this memory of a dead man's skin. Had Evan and Margaret known that death and coal would follow them to their new country, I wonder if they would have still crossed the ocean on the *SS Wyoming* and made their way across the plains to join the Saints.

The Thomas family boarded a train in the East, traveling west on a railroad that had been first invented in their hometown to transport coal. The iron of the tracks that first canvassed their new world was forged with the coke fired from Welsh coal, coal that had passed through the hands of Thomases for three generations. The tracks civilized the prairie and mountain lands, bringing manufactured goods to the pioneers and farmers who made their own soap with white ashes and animal fat, spun and knit their own clothing, and grew their own food. With the expanding railroads, the capacity for coal extraction increased, further fueling development and the power to create goods and move them—setting in motion a cycle of consumerism and fossil fuel that defines much of the way we live today.

With their five children and Evan's father Frederick, my great-great-grandparents settled in Logan, Utah, a small community north of Salt Lake in a valley in the Wasatch Range of the Rockies, below pine-covered mountain slopes where snowmelt trickled down during the summer months. Here my great-grandfather Zephaniah was born, the second American to join this poor Welsh immigrant family.

Evan and Margaret were determined to make a break from their past in coal, so Evan found work on the railroad, a job that offered the pleasures of light and air, but otherwise was not a far cry from the back-breaking labor of the mines. One day, after working for twenty-four hours straight, Evan was ordered back on the tracks by his supervisor. Exhibiting the fiery temper he was known for, Evan refused and was fired on the spot.

Evan had heard of the excellent wages paid to experienced colliers by

Courtesy of the Western Mining and Railroad Museum

The town of Scofield

the Pleasant Valley Coal Company that mined in Emery County. By now, Evan and Margaret had seven children, and in order to feed his family, Evan was forced back into the trade he knew best. The Thomas family moved two hundred miles south to Winter Quarters, a coal mining camp above the city of Scofield.

At the time, Scofield housed 1,800 residents and the business district was a mile long, sizable for a place that was such a far reach into the mountains. It was a coal town, like the one the Thomases had left behind in Wales—the men who left the mines at the end of the day continued their comradeship in seventeen busy saloons. An English gentleman complained of Welsh miners that "the workmen of this black labour observe all abolished holy days and cannot be weaned from that folly," and a similar complaint could have been made about the miners in Winter Quarters. The young women whose beaus and husbands spent most of their daytime hours underground looked for every excuse to plan parties. According a local resident from that time: "People would get out and make their own good times. They had dances in the meetinghouse, for instance. The dances were usually on Fridays because on Saturdays they made them stop at midnight. The next day was Sunday and there was always a squabble about closing up the dance."

Much of the community's social life surrounded the local chapels—the Latter-day Saints meetinghouse was prominent in Winter Quarters. Due to the numerous Welshman employed at the mine, singing was a common evening time diversion, and miners and their families would gather to sing at the homes of neighbors who owned instruments.

A piano was rare. Most of the residents of Winter Quarters had few possessions: their one-room cabins were equipped with a stove, beds, a table, and chairs. Some families had rockers, and occasionally a phonograph. Still, a miner's home would have been set off a few paces from the neighbors. Americans, with their characteristic love of space, made sure that even coal villages had freestanding houses. They were small and unelaborated, but had their own grounds, a mark of status in Wales, where most miners were packed into company owned blocks of terraced housing. In Scofield and Winter Quarters, most families built their houses with sawmill-cut boards placed vertically, topped by a peaked roof. Though privately owned, these houses were built on company land and families paid a monthly land rent.

The women took great pride in their homes. Contrasted with the atmosphere of the coalfields—a masculine landscape based on muscle, waste, and the triumph of mechanization over wilderness—the culture of the home was one of order and cleanliness. Most mining towns, though initially inhabited by single males and enough public houses and saloons to serve them, would eventually include enough women to balance out the population.

Margaret would have had a very similar life to that of her mother Ann in Wales. Fighting back the dust of the coal mines, she would spend many hours scrubbing her wooden floors white and stirring the clothes of her men in a big copper pot of boiling water to release the body and earth grime staining the cloth. Aside from preparing meals, chores like these would occupy her days, wearing her hands as rough as her husband's. Margaret's rejection of Mr. Lewis's son, her voyage across the ocean, and renunciation of her former life would come to mean only one thing: at night the man she loved would emerge from the mine and they would gather with their children to read the Book of Mormon in a community of Saints in hills more commanding and remote than those she had left behind.

Winter Quarters was a relatively new coal venture. In 1870, farmers had begun to scoop coal from the surface into wagons and cart it forty-five miles through Spanish Fork Canyon to the larger town of Provo, a four to five day trip over rough mountain roads. In 1875, the Pleasant Valley Coal Company began to invest in the area—the first miners dubbing it Winter Quarters, due to the snows that came early and trapped them in the coal pit until February. The Calico Road (a length of rail built by trading labor for yards of cloth) later connected Winter Quarters with the Denver and Rio Grande Railroad, preparing the area to produce half of Utah's coal output by 1899.

Chinese laborers who had originally worked on railroad gangs were hired to mine in Winter Quarters. Their skill in perfectly proportioned tunneling was admired in other mines, and they placed small demands on the company, as they were willing to make their homes in the tunnels they dug. Miners of other ethnicities, who had just begun to unionize, saw their leverage with the company slipping and resolved to settle the matter with the violence characteristic of mining towns. According to an oral history from 1890, a group of miners, reportedly mostly Welsh, roughed the Chinese into a railroad car, sending them with a push down the incline from Winter Quarters to the valleys below. Ten miles down the tracks the car turned over. The Chinese scattered, never to mine coal in Castle Valley again. Thankfully, this was three to four years before the Thomas family had arrived in town, and Evan and his boys were not involved.

Racial tensions continued at Winter Quarters. Faced with a shortage of labor, the company imported a sizable number of Finns. The miners from the British Isles and France regarded these new immigrants as nonwhite due to their perceived foreignness and their superstitions, so they were relegated to rows of houses farther up the canyon, which came to be known as Finn Town. A subsequent migration of Greeks settled near them in 1903. It was in Finn Town that a murmur began late in 1899. Miners began to speak in low voices about "bad spirits" near the Winter Quarters mines.

At the turn of the century, Americans in general had great faith in their frontier and their factories. In the coal industry in particular, miners were paid better than any other workingmen, and the national market had never been better. The year previously, Utah had not had any accidents and every mine had met regulations. As a result, the fears of the Finnish miners were dismissed by the others as the superstitions of a backwards people.

On May 1, 1900, Evan and his three boys awoke at five o'clock in the morning to put on warm clothing. Even in the spring and summer, the interior of the mountain was cold and wet. Likely, their lunch was packed in tin buckets by Margaret, who wiped the sleep from her eyes, and then kissed them and sent them off to the mine, glad to have some peace for Dewey Day preparations. Two years earlier, General Dewey had defeated the Spanish fleet in the Philippines in five hours, without loss of man or ship. This victory fed into America's sense of manifest destiny, and the repercussions were felt even in this small mountain town. Winter Quarters had just received a contract for two thousand tons of coal a day from the US Navy. Miners were only required to work a half day on this holiday, and later that evening they would celebrate with a party, a dance, and Dewey Cakes at the church house.

My great-grandfather Zephaniah, or Zeph as he was known, woke with his older brothers Frederick and Evan Jr. to work with his father. He had been allowed to graduate from primary school, but after sixth grade he was expected to help earn his keep. In 1900, he was seventeen years old and had been working in the mines for five years. The already sizable Thomas family had expanded by another member since the move to Winter Quarters, and multiple sources of income were doubtless a necessity to feed a Welsh family of eleven.

The workday would start with the fireboss, who carried a torch by hand through the tunnels to burn the mine clear of flammable gases that had accumulated overnight. After that, the foreman would assign each miner to a room. The men would then turn to the mouth of the mine and walk or take a mantrip—a line of cars pulled by horses or mules—yards or even miles into the mountain to their designated spots, where they would lay down their tools. Their path would be lit by the oil-powered lanterns attached to their soft cloth hats. The open flame, rising out of an oil lamp shaped like a small teapot, extended six to eight inches above the brim, flickering off the roughly cut, black walls of the tunnels. That day Frederick and Evan Jr. split off to go to their assignments in Mine No. 1, but Evan Sr., feeling he had been given an inferior room, "rose to his full five foot three inches and roundly denounced the foreman, the superintendent, and the Pleasant Valley Coal Company generally before taking his youngest working son and heading home."

Methods of mining had progressed with new technology, and although miners still swung picks, blasting powder accelerated the process of extraction. Once reaching his room, a miner would shore up the roof with wooden poles. Then he would proceed in one of two fashions of removing coal from the face. Undercutting required a man to lie on his side and cut the coal from the bottom with his pick, making it easier to remove the *burden*, or higher coal, later. This was the older, slower way, used only when the floor of a mine needed to be level. Shooting was messier but faster. With a hand-turned drill, a miner would bore two to four six-foot-deep holes into a coal face. The next step was preparing the charge, which, according to a miner who worked as a boy in Winter Quarters, went something like this: "First they would make a paper cartridge from butcher paper or wrapping paper. They had a round stick, like a broom handle, that they would wrap the paper onto. Then they would fold the paper over the end and pound it to block off one end. Into the open end they would pour the powder. The charge would be eighteen or twenty inches long and would contain a pound or a pound and a half of black gunpowder."

This blasting powder was poured from a twenty-five pound wooden keg that had been punctured with a pick and then stopped up with scraps of paper. The homemade charge was popped into one of the holes, followed by a copper needle that would hold the place of the fuse, or *squib,* which was inserted later. Next, clay was tamped around the device. If this mud was not tamped in well, holding the charge of blasting powder deep in the face, the charge could backfire and ignite the coal dust.

Before a man lit his squibs, he would alert his comrades in the rooms on either side of him. Touching each tip with flame, he would duck into a safe corner with his hands over his ears until the shots went off. Immediately, he would cough his way back to the cracked surface of coal, batting the dust from his face, and pry out the chunks with the tip of his pick. Miners were paid by the weight of lump coal they extracted a day. They did not have time to wait for the dust to settle.

In Mine No. 4 at 10:28 a.m., the exact hour later discovered on the watch of a dead Finn, miner Bill Boweter overheard two miners in a room adjacent to him say they were going to blow some shots. They had overestimated the thickness of the wall that divided their room from the next, and the ensuing blast brought the entire wall down. The coal dust exploded into a surge of white flame, which lifted Bill off his feet and threw him against the wall. The cloud of coal dust billowed and burned, orbs of fire rotating in the middle of an onslaught that thundered from room to room, igniting each miner's powder kegs with bangs and bursts of light.

The local justice of the peace, Joseph Saunders Thomas, and his son Joe, who grudgingly held the drill steady on the coal face for his father, felt the shock waves before they saw the conflagration. Dropping their work, they ran for the entrance, Joseph grasping his son seconds before they were picked by the energy of the blast and flung. John Wilson stood near the mouth of Mine No. 4 with his mule. The explosion threw them 820 feet. The mule was killed by the impact, but John's fall was halted by a tree branch that impaled him right through the middle.

Later, a train sent from Salt Lake City to convey the seriously wounded carried John and four others to the hospital. After several operations, John returned to Scofield to live out the remainder of the seventy years of his life, never setting foot in the mine again. Harry Taylor, also aboard the train, was stopped by a reporter as he was being moved by stretcher into Saint Mark's Hospital. In the words of the May 2, 1900, *Salt Lake Tribune,* "his face [was] swollen and bandaged until there was barely any room open for his mouth and eyes," and, although Harry sustained a number of injuries caused by rocks and debris, he "believed he came out comparatively in good luck." The journalist scribbled down phrases at a time, as Harry would pause to summon his courage between severe attacks of pain. According to Harry: "I was repairing some track out on the dump when I started toward the mouth of the tunnel to get some more tools. I got about fifty feet away from the tunnel mouth when all of a sudden there was an awful report and at the same time a black cloud filled with rocks bore down on me like a streak of greased lightning. ..." And then, with what the journalist called a ghastly attempt at smiling, Harry added: "Next thing I knew I woke up with a man pouring brandy down my throat." In terms of his good luck, Harry understood the matter right. Most of the men who were not killed by the blast soon succumbed to poisonous gases.

The town did not react immediately; it was Dewey Day and some thought the boom came from fireworks being shot off early. Evan had undoubtedly returned home to repeat his tirade, rising up on his toes and shaking his finger, adding additional color to his denouncements to impress his wife. Likely, Margaret shook her head, peered at him knowingly with her deep blue eyes, and hoped his boss would take him back tomorrow. At the sound of explosions, Evan must have paused in his cursing and Margaret in her kneading of dough. Was it fireworks? Or—a sharp intake of the breath and a rush of terror—the mine? I imagine Zeph straightening up from where he sat, safely slumped in the corner with nothing to do now that he had been pulled from both work and school. Both Evan Jr. and Frederick were still in the mine. The family must have run from their home, the air just beginning to reek of coal dust as a cloud of smoke rose from the hillside.

Men who outran the explosion went back into the mine for their comrades. At this point, rescue crews had not been fitted with protective equipment, so a man would walk as far in as he could until he dropped. Others behind him would put their hats to their mouths, dash forward, grab hold of his feet, and drag him out. This is how they tested the poison content of the air, and progress into the depths of the mine was made in increments. Most of the men pulled from Mine No. 4 were already dead; the few that showed signs of life expired soon after being brought to the air. All were burnt badly—"to a cinder," reported a local schoolgirl.

Rescue teams were later assembled, wearing helmets that resembled those of early astronaut suits: two thick tusks emerged from the helmet and connected to an oxygen tank strapped to the front. Once the teams had penetrated through to the far reaches of Mine No. 4, they moved on to Mine No. 1, which the afterdamp (carbon monoxide) had filled earlier, making it impossible for unprotected rescuers to enter. Progressing slowly, the men moved in a line along the back corridor downhill from Mine No. 4 to Mine No. 1. In the first room, no larger than the size of a frontier living room, thirty-five men lay dead. They had fallen in positions revealing their scramble toward fresh air.

One rescuer recounted his experiences over the two days following the accident:

> Going in we saw a number of dead, but our object was to find any that were alive first ... We only found a few alive and one of them has since died. Number

Four was so blocked that progress was slow and very dangerous and we had to carry the men out on stretchers, as cars could not be used. A good many in Number Four were badly burned and mutilated.

After working a while in Number Four we worked our way into Number One where nearly all the men who died were asphyxiated. A great many people have asked if the men who were killed by the damp suffered much. I can say that they did not, and know that to be the case, because I have gone through the experience to the stage of unconsciousness during the last two days. Many of us in the rescue parties were overcome by the damp and were carried back into the purer air by our companions. A whiff of this damp almost paralyzes a man, and a good breath of it renders him unconscious. Then he falls as if in a sleep and dies unless instantly carried to the purer air. What struggles take place after that first breath are the struggles that nature puts forth automatically. I have seen our men fall and struggle but they knew nothing of it.

We found the dead in every conceivable attitude. One man had filled his pipe and sat down to light it. The damp struck him and he died then and there, with the filled pipe in his outstretched hand ... We found men in groups who had evidently sat down to consult. Other groups had been overtaken as they rushed ahead of the damp.

How can we go through the ordeal of picking up the dead shift after shift? Well, I've thought of that, too. We made up our minds not to give way to our feelings; to stifle them and ignore our thoughts of everything except the work at hand, otherwise we could not do a thing. Well, when I have gone through there and turned over a man who was a friend and intimate associate, perhaps of a year's standing, the sentiments of grief stirred in all of my being, but I repressed them. It was either that, or else drop down by the side of my chum, take hold of his cold hand, and just cry my heart out, for we have hearts just the same as the rest.

Some of our men were nervous at first, for the scenes, in spite of all resolutions, did excite and move us; but when the death list grew to fifty, then to a hundred, then to a hundred and fifty, and now to nearly two hundred, all got over it.

Frederick and Evan Jr. were working in Mine No. 1. One of the survivors, W. U. Winston, indicated that he heard "a rumbling noise in the distance" and felt "a wave that can hardly be described except to all who have been in explosions." He dashed to the mouth of the mine, convincing others to join him as he ran, leading nine of his comrades to safety. They nearly suffocated from the afterdamp that reached them three to four minutes before they exited the mine. Thomas Pugh, a fifteen-year-old boy, heard the noise and put his cap in his teeth to cover his mouth and nose, sprinting a mile and a half to the entrance. One hundred and three miners escaped the afterdamp in Mine No. 1. Others were not killed immediately and hoped to be rescued. One man and his son held out until four hours after the blast, leaving the following note to his wife: "Ellen, darling, goodbye for us both. Elbert [his son] said the Lord has saved him. We are all praying for air to support us, but it is getting so bad without any air. Ellen I want you to live right and come to heaven. Raise the children the best you can. Oh how I wish to be with you, goodbye. Bury me and Elbert in the same grave by little Eddy. Goodbye Ellen, goodbye Lily, goodbye Jemmie goodbye Horace. Is 25 minutes after two. There is a few of us alive yet. Oh god for one more breath Ellen remember me for as long as you live. Jake and Elbert."

Evan Jr. was working near the entrance and could have joined the miners escaping to safety, but he turned back into the mine to find his younger brother. The rescuers found Evan and Margaret's two dead sons hand in hand. Evan Jr. was twenty-five and Frederick was nineteen. Forty years later, Zephaniah could still barely speak of the incident. Frederick was just two years older than him and was his idol.

During rescue efforts, it had begun to drizzle. Hundreds of women and children gathered around the mouth of the mine, moaning and crying, trying to catch glimpses of the bodies carried out in burlap sacks and piled in the mantrip. The rescuers carried the dead miners to a storeroom, and then to a boarding house where Clarence Nix, a quiet and shy man of twenty-three who was the manager of the company store, tagged each one, identifying the men to whom he had sold blasting powder, tools, and canned goods every day over the past several months. Then, to spare the feelings of the women, these rough men carefully removed the battered and burned clothing from their friends. With sponges soaked in tubs, they washed soot and blood from the mutilated bodies, and then rolled them in old quilts. The dead miners, now ready for burial, were carried to the schoolhouse or the meetinghouse, where they would be taken home by their families.

The May 2 *Salt Lake Tribune* ran the title: "MOST APPALLING MINE HORROR: Greatest Calamity in the History of Mining in the West. EXPLOSION AT SCOFIELD KILLS 250." This original estimate is closest to the 246 counted by the men carrying the bodies out of the mine, but 200 is the number settled on by the company. Even after the final count, the Finns maintained that fifteen of their men were never found. Since many of the Finns were single men, separated from their families by thousands of miles, there is no way to confirm or reject this assertion. These men were so unassimilated into their new environment that it is possible that some still lay buried in the Winter Quarters mine, unaccounted for by the company or their coworkers.

The Winter Quarters mine explosion topped the charts as the most devastating mining accident of its time, and it currently ranks fourth in US history. The detailed accounts of rescue operations and the suffering of the people struck a sympathetic vein in the surrounding communities, and even throughout the world. The *Salt Lake Tribune* printed condolences from President McKinley: "I desire to express my intense sorrow upon learning the terrible calamity, which has occurred at Scofield." The same issue also printed a letter from President Loubet of France. London's *Daily Telegraph* reported: "There will be deeper sympathy with America in this awful catastrophe than has been evoked by any event of the other side of the Atlantic since the loss of the *Maine*."

At the time, there was no system of government sponsored welfare, and the explosion had left 107 widows and 268 fatherless children. The first relief came by way of schoolchildren from Salt Lake City, who went door to door gathering flowers for the funerals. Many of them were not hothouse flowers, but ones people had cut from their gardens: bunches of lilacs, pansies, and violets. These were loaded into the baggage car of a train and bound into bouquets on the trip to Scofield by the women who traveled with them. In addition to burial clothes, the company had provided coffins for the bereaved families at the cost of ninety dollars apiece—only 125 were available in the whole state of Utah at the time, so 75 had to be shipped from Denver.

Four days after the explosion, the day of the miners' burial was overcast, with scattered rain and a stiff wind. A photograph shows two Finns in shiny black caskets, lined in shimmery white cloth, with bouquets of flowers scattered over them. The face of one is half covered, and one arm, doubtless scarred, is wrapped in linen. The man in the casket beside him appears unmarred. His eyes are closed over a short nose and an enormous black handlebar mustache. His arms cross his chest, the hands displayed in handsome white gloves. In death, he appears to have had a much more genteel occupation in life then wielding a miner's pick.

Fifty volunteer gravediggers from Provo were hard at work preparing the graves, some as little as three feet apart to accommodate the many burials. As the wagons carried the caskets to the graveyard, flowers were handed out from the railroad cars, with some bunches placed on the coffins and others given to the children and women. Reverend A. Granholm, a Finnish Lutheran minister from Wyoming, honored the sixty-two Finnish miners who died in the accident with a service conducted in his native tongue. An interfaith funeral was led by three Mormon apostles. An image from the later internments shows a large crowd of mostly men in the act of removing their hats. Piles of earth are stacked up between the grave markers, cut out of boards by the local sawmill operator and labeled with the names of the dead written in lead pencil. In the foreground a boy leans against a heap of dirt, tilting his head for a clear view of the man being lowered into the ground. I think of Zeph, watching his brothers buried in soil turned from the mountain that killed them. He was a quiet man with measured emotions when he grew up, but as a boy, looking down into

Courtesy of Western Mining and Railroad Museum

Burial services at Scofield cemetery

those graves dug side by side six feet down, Zeph must have seen little hope in his future.

There wasn't a family in town without a personal connection to the tragedy. An eighteen-year-old bride of not many months lost her father, her two brothers, and her husband. The Hunters lost ten men, all the male members of the family except two. The Louma family's story is particularly terrible. Seven sons had left Finland to make their fortunes in America, finally settling deep in the Wasatch Range to mine coal at Winter Quarters. They sent for their elderly parents, writing that they could earn enough money in America so that Abe Louma and his wife would never have to work again. Three months before the tragedy, the two arrived in Scofield and began to adjust to life in a new country, surrounded by their children and grandchildren in the small miner's house in Finn Town. Six of their sons and three of their grandsons were killed in the explosion. After burying his posterity, Abe Louma told his wife: "If I don't live longer than a cat, I am not dying in America." The Loumas rode the train back through the mountain pass, escorted by their only living son. From California, they returned to Finland by ship, repeating a journey they had undertaken months earlier with anticipation.

Due to the speed with which funeral arrangements were made, some miners were buried under the wrong name. John Pittman was killed in the blast and presumed to have been buried in the graveyard with his comrades. A month after the explosion his son dreamed of an angel who told him his father's body was still in the mine. The next day the son reported this to his Mormon bishop, who didn't make much of the boy's claim until he himself had a dream that night. In his mind he moved through the rooms and corridors of Mine No. 4 looking for John's body, until he found a rock where three little girls dressed in white were laying flowers overtop. The next day the bishop gathered a small crew of men and followed the path indicated in his dream. After digging through a hundred feet of fallen rock, he found the poorly preserved body of John Pittman, who was unmistakably identified due to the bright red socks, still on his feet, that had been knitted for him by his wife. The man initially buried under Pittman's headstone was never identified.

The day after the accident, a reporter called upon the residents of Salt Lake City to offer their assistance: "There are in the city hundreds today who have passed their summer vacation within a stone's throw of where the accident occurred, and they cannot but have a pang this morning when they think of the children who were so fond of the strangers that passed near their humble homes being robbed of the breadwinners of the household and thrown upon the charity of the world. There are few if any families, it is safe to say, who have much to the good, while it is more than probable that their accounts are already overdrawn at the store."

Catholic Bishop Lawrence Scanlan offered to take fatherless children into Saint Ann's Orphanage in Salt Lake City. When a shortage of food was reported, Salt Lake bakeries donated three thousand loaves of bread and the millers' association sent four hundred sacks of flour. Others gathered butter and eggs to be sent, and members of various communities around Utah went door to door to gather funds. The miners were the best-paid workmen in the state, but the poorer farmers did their part to relieve the suffering of the victims of the disaster. A man from American Fork sent a dollar folded into a letter, apologizing for his meager donation—he'd had to borrow it.

The Pleasant Valley Coal Company, in addition to paying for the funerals, forgave the debts of the bereaved families at the company store and gave $500 to families for each miner that was killed. The Colorado Fuel and Iron Company donated $5,000, and Levi-Strauss sent $250. The LDS Church contributed $2,500 and also held a concert at the Salt Lake Tabernacle, where the proceeds of 10,000 tickets were donated to the community. Other towns sold tickets to baseball games, and the Skelton Publishing Company offered to publish a book about the explosion, which James W. Dilley, the local schoolmaster, agreed to write.

In total, $118,000 was collected. Because there were many opinions about how the money should be distributed, the governor assembled a committee of five members to assess the needs of the community in order to divide the funds. Their final decision was as follows: "Assuming a pension of twenty dollars per month for three years, each widow over fifty will get $720 dollars. Widows under fifty will receive $576 dollars. For each boy under fourteen and each girl under fifteen; $108 dollars which will go to their mothers or grandparents. Each child with both parents dead will get $432 dollars. A parent of a deceased miner, who depended upon his son for support will get $720 dollars. If both mother and father are alive and fully dependent upon their son's earnings, they will receive $1,080 dollars. If a parent was getting some kind of partial support already, say, from another son, the parent would get $900 dollars. If just one parent was alive, $540 dollars."

The committee allotted fewer dollars to the younger widows on the assumption that they were more likely to remarry. Relief would last for three years, and by then the widows would have to find some way of supporting themselves. Many used the money to move down into the valleys in farming communities where the winters were milder. A significant number of the surviving miners left, but the majority of the Finns, who had never really become integrated into society, stayed. Perhaps the cooler temperatures reminded them of their homeland.

Evan and Margaret received $2,240 in payment for the deaths of two sons. Instead of using it to spare the rest of their children from work in the mines, they erected a monument. An eight foot marble obelisk inscribed THOMAS stands above the resting place of Evan Jr. and Frederick in the Scofield graveyard. Its size suggests their parents were more consumed with the grief of the moment than with the practical challenges of the life that would follow. Perhaps the thought of buying flour with funds procured by the death of their boys was too much for Evan and Margaret—investing in such a grave marker might have seemed the only way to dispose of it properly. Work to put the mine back into operation commenced a week after the accident. The miners who stayed and a number of new recruits spent long hours clearing out the debris and shoring up the roof. I can only imagine the feelings of Margaret, who watched Evan leave every morning with the awful memory of a rumble not far from her thoughts. Evan walked to the mine on these dreary mornings alone; after the accident, Zeph did not follow behind him.

7

Zeph and Maud

According to family lore, Zeph bought his first pair of Levi's at age seventeen from a clerk named James Cash Penney at the Golden Rule store in Evanston, Wyoming. (Two years later, Penney opened what would become the flagship J.C. Penney store in Kemmerer, Wyoming.) How Zeph had ended up here from that narrow canyon in Carbon County, Utah, I will never know. Perhaps he caught a train in Scofield and headed directly to Wyoming, or perhaps he arrived there on the twisted path of the wanderer he was bent on becoming. In Wyoming, he found a job in a mine, the only job for which he had any training. This purchase of jeans was his first step into a manhood where he worked his own rooms and drew his own pay.

Wyoming was only the first stop in Zeph's period of *tramp mining*—a term applied to temporary workers in the early twentieth century who would move from state to state, and from mine to mine. They would earn a makeshift living, camping in shacks and descending down a particular mine shaft until "elsewhere" called. After a year in Wyoming, Zeph moved north to Washington State, where he worked seams of coal only thirty inches high, lying on his side the way his Welsh predecessors had done. He lived in a boardinghouse run by landladies who must have taken to the quiet and polite eighteen-year-old with a bit of tragedy behind the blue eyes he had inherited from his mother. Every day they asked what he would like in his lunch. Pie or cake was always his response. They never tired of spoiling him, or he of eating dessert, a hankering he would pass on to his son Bobby, my grandpa.

Coal mining had become a major American endeavor. By the turn of the century, with a seemingly inexhaustible amount of inhospitable mountain ranges, poor immigrants, and enterprising spirits, the United States was extracting more than one hundred million short tons of coal a year over the amount produced in the British Isles. Most mining was taking place along the Appalachians and in the Midwest. Pennsylvanian miners produced more than all the rest of the states combined to feed the factories in the East. In the West, mining had much to do with the symbiotic relationship between coal and rail. Tracks were laid to reach mines, and mines extracted steam coal for the trains, each justifying the existence of the other. Colorado became a significant producer of coal, but instead of seeking work in more established mines, Zeph continued his pursuit of a still more distant and remote coalfield.

Coal was mined in the stretch of the Rockies between the provinces of British Colombia and Alberta from 1880 to 1950 until petroleum was discovered (now Canada's coal is largely imported from the United States). The industry reached its height before the First World War, and this is when Zeph crossed the border to work in the Canadian coal mines. There is considerable distance between the Rockies in Utah and the Rockies in Alberta, and Zeph communicated with his mother by letter. On May 22, 1902, Mine No. 2 exploded in Coal Creek, Alberta, killing 128. It had been just over two years since Margaret had lost her two oldest sons, and her memory of waiting in the cold May rain for Evan Jr. and Frederick to be carried out of the Winter Quarters mine would have been fresh. Zeph was not involved in the Coal Creek accident, but he sent no word to his mother for months.

During these years that Zeph bounced from mine to mine across state and national borders, he took up smoking cigars and drinking coffee, two habits forbidden by the LDS religion, although he later discarded them before his marriage. Zeph stayed away from more destructive pursuits—the saloons and the women his fellow miners frequented—but his cups of coffee and cigars might have represented the rebellion behind all his drifting. Years later, when my grandfather Robert had to give up college for a while, Zeph told him that he had always wanted to be a doctor. Possibly, as Zeph left Scofield, he dreamed of a distant coalfield in which he would have the break he needed to pay his way through school.

Before he turned twenty, Zephaniah returned to his family and mined alongside his brothers. They had pooled their money to buy a farm for their father in Cleveland, Utah, which was in Emery County, bordering Carbon County. After a life underground, Evan would be a farmer. A miner's eye for topsoil was evidently not equal to that of a man who had worked all his life on the land; once irrigated, Evan's farm coughed up alkali. Another attempt at departing from the family occupation had ended in failure. The farm that could not support a profitable crop at least allowed them to grow orchards and raise hens and bees. The move seemed to suit Margaret. She was cherished not only for her role as the nurse alongside the town doctor when a flu epidemic broke out in Cleveland in 1918, but also for her constant acts of kindness. Baskets were filled with fruit, eggs, honey, and baked goods and sent out on the arm of her granddaughter for the elderly and lonely in their community, the farm providing enough at least for Margaret's charity. She died on her land at age seventy-six, slumped under the crab apple tree with an apron full of apples for jelly.

Zephaniah inherited some of his mother's disposition and was usually gentle. He never was hot tempered like his father, but he could be stern in his own way. He wouldn't tolerate the drunkenness of some of his brothers, who accepted the lot of their lives more recklessly. He was the most respected member among his family, careful in his behavior and also his dress. Light on his feet on the dance floor and easy on the eyes, at twenty-one he was the most sought-after bachelor in Cleveland.

Lily Maud Rencher, after finishing her normal degree at BYU Academy, taught school for a year in St. George and then moved to the small town of Cleveland to educate the children of farmers and miners. Miss Rencher—called Maud—was a large woman. Her Swedish, Irish, and Welsh background had not blessed her with beauty. At school she had been a student body officer, and in the words of one of her professors, expressed to my grandfather later: "She was kind of—a big girl, but wow could she speak." Maud took the train from Provo, her few dresses packed in a suitcase. Provo was a small town at the time, and Maud had grown up in a farming community. Even so, there must have just been a hint of romanticism in her mind as she struck out into the wild because in the undeveloped West at that time, mining communities were wild. Confident in her charms, she knew she would endear herself to Cleveland, but perhaps not to its most eligible man, whom she fixed her eye on one night at the one of the local dances.

"Who's the snappy-looking man in the black sateen shirt?" she asked her friends as they clustered in a corner, sizing up the room, pinching their cheeks for color, and smoothing their dresses.

"Oh, that's Zeph Thomas. Everybody's looking to catch him. He's the oldest and most eligible bachelor in town. You don't have a chance with him."

Maud reveled in challenges and told her friends that he would walk her home that night, sealing the deal with a bet.

"I know something about you." These were Maud's first words to Zeph, who had probably only taken polite notice of the new schoolteacher until her bold words jolted him to attention. Zeph was 5'7" tall; she was two inches shorter and doubtless a bit wider. She also had freckles—something my grandfather once told me, noticing the sprinkling of little brown dots on my five-year-old nose.

Zeph and Maud Thomas circa 1920

"What do you know about me?" Zeph asked, surely leaning forward a little and peering at her.

"Oh, something rather interesting," she teased.

The rest of the night Zeph followed her around, trying to pry out her secret, one she must have told him years later: Maud had decided that Zeph would fall in love with her that night. She was her father's favorite and was accustomed to being circled about by listeners Once he heard her speak, she knew Zeph would not forget her easily. Maud taught in Cleveland for three years, during which there is no record of Zeph and Maud's courtship. A prolific letter writer, Maud must have shared the details with her friends and her family. None of these letters have survived, and for some unknown reason, Maud and Zeph never shared these experiences with their posterity. Three years after Maud and Zeph first danced, Maud returned to BYU alone for a course in public speaking and then accepted a post in Bunkerville, Nevada, over three hundred miles away from Cleveland.

Here, in a small outpost of the descendants of LDS pioneers, Maud taught school for another three years. The summers were so hot that the whole town would leave their homes and shops and lie in the irrigation ditches to cool off. Under the dry sun of Bunkerville, Maud fell in love with Joseph Walker, a coworker who was preparing to attend medical school. They became engaged, but Maud finally broke it off because she did not believe he was spiritually minded. At the age of thirty, she returned to Utah. Now twenty-eight, Zeph worked in the Sunnyside mine as a boss driver, supervising the movement of coal from room to room as it was shuttled along the tracks by the mules. During Maud's three-year absence, Zeph and Maud must have kept in touch, developing the relationship they had begun seven years previously at a community dance. When Maud arrived, Zeph met her in Salt Lake City, and they married on June 10, 1911.

After their wedding, Maud joined Zeph in Sunnyside, where, in addition to the mining operation, there were 713 coke ovens. These were built of brick, shared sidewalls, and stretched out in long lines. A railroad car above them would dump eight tons of coal into each oven through an opening in the top. In an airless chamber at one thousand degrees Celsius, the coal would heat until all of its smoke, tar, and water was burned out, leaving a gray and porous substance ideal for smelting iron. At this point, the coke was ready to be pulled. The operator of the oven would unbrick a door the size of a kitchen cabinet, spray the coke with water, and drag it out load by load, finally transporting it by wheelbarrow to the trains. It was a hot and tedious job. An experienced worker only pulled two ovens of coke per day.

In Sunnyside, Maud gave birth to William, who was followed by Robert in 1918. My grandfather had only vague memories of the coke ovens of his birthplace—his mind's images no doubt clouded by the smoke from lines of tens upon hundreds.

8

The Castle Gate

The section of US Route 6 that passes through Price Canyon is one of the most dangerous highways in America. Semi drivers on their way east lumber along this highway that curves through the mountains, and at night only stars and headlights illuminate the way. On many days in winter it is better to avoid it. Even the elaborate winter snowplow system in Utah is insufficient to remove the danger of ice that accumulates at Soldier Summit, the road's high point. From Provo to Price, the drive is seventy-four miles, a journey to my parent's home that I never looked forward to.

While I was finishing my education at Brigham Young University, my father had been offered the presidency of the College of Eastern Utah—a small college that serves Price and the surrounding towns, largely populated by the descendants of coal miners from Carbon County. The coal mining industry has gradually been replaced by other means of energy production, although there are still members of these communities who mine or drive coal trucks. The college and the public school system provide most of the jobs in Price. In his formal acceptance statement, my father mentioned that his father and grandfather had worked in the mines in Carbon County, a fact that largely influenced his decision to move with my mother Ann and my younger siblings to a home overlooking the college and surrounded by tawny colored cliffs, flat-topped and sandy.

Despite the dryness of this landscape, central Utah has a particular sort of loveliness. Even now, I dream of a time when I can come "home" to the mountains there and spend a summer exploring with a backpack full of brushes and oil paint and a canvas under my arm. With a boulder for a stool, I would squint my eyes to probe out the color that lurked over bare rock, the hues changing according to the time of day. Near the mouth

Courtesy of the Western Mining and Railroad Museum

The Castle Gate

of Route 6, the tall peaks covered with pine look violet in their gullies at dusk. Farther south, the mountains are remarkable in their treelessness, covered with short brush. Here the dark sandstone bluffs are carved into shapes that, when one squints, take on the character of droopy-faced hill trolls. Past the level plain of the Price River that winds in wide meanders, the Castle Gate reigns, marking eight miles to Price. The *Guide Book of the Western United States*, commissioned by the US Department of the Interior in 1922, describes this structure as a "peculiar gatelike passage 2 miles above the town, the sides of which seem to be walls or dikes of sandstone projecting from the sides of the canyon. When viewed from a point directly opposite it the rock wall on the right looks like a thin finger, but when seen from a point farther up the canyon the walls on the two sides seem to project so far into the canyon as almost to obstruct it and to bar the railroad from further progress."

In 1922, one wall of the Castle Gate stood 500 feet tall, and the other 450 feet, the two nearly equal in length. On the path of the old railroad, the two sides would converge from a distance and open as the engine huffed around the bend. In 1966, one side of the gate was destroyed to make way for Route 6. Where the 450 foot gate once stood, the side of the mountain has been shaved off and tapered, but the 500 foot wall still towers over the road, solitary and magnificent. When I was a four-year-old child waking up from a nap on a drive to central Utah, my mother told me, "We are in Price Canyon." I thought she said *rice* and was confused because the canyon did not look like rice, but I remember distinctly that last remaining gate, a landmark also inscribed in my grandfather's childhood memory.

Not far off Route 6, deeper into the canyon, are the remains of the town of Castle Gate—a graveyard with several hundred occupants. No Thomases are buried there, but it is the resting place of many of Zeph and Maud's friends. There are also likely many of my Grandpa Thomas's kindergarten classmates buried in that cemetery, the boys who followed their fathers into the mines and the girls who married them. In 1978, the town was dismantled and the remaining houses moved to Helper, three miles down the canyon. Now the Utah Castle Gate Power Station is all that remains of coal operations there. The plant is set off in a canyon below a curve in Route 6, where the road cuts through a tall wall of sandstone. Every time I whirl around the bend I note the three foot thick coal seam in the face, making one magnificent black stripe in the surrounding tawny colored rock. At night the smoke of industry clouds the lights of the power plant, an effect that is almost dreamlike, a mirage in the thick darkness. Here the road becomes more precarious, narrowing to one lane with no passing. Being stuck behind a semi guarantees adding an extra ten to twenty minutes onto the trip to Price.

It was just three years after the Winter Quarters mines opened in 1888 that the shaft of Castle Gate Coal Mine No. 1 was sunk. This mine took preeminence over the other mining operations in Utah when a seam twenty feet thick was discovered in the newly tunneled Castle Gate No. 2. This coal was determined to be the best in the region. In an old photograph, tall metal columns hold up the ceiling of a coal face that dwarfs the miners below, who beam into the darkness with their carbide lamps. Almost ten years later, in 1897, Castle Gate entered the history of cowboy lore as the site of one of the most daring holdups in the American West.

One hundred miners from the Pleasant Valley Coal Company waited for their salary around the tracks of the Castle Gate money train, among them fourteen-year-old Zeph Thomas, whose family was currently working

Courtesy of the Western Mining and Railroad Museum

Castle Gate miners handrilling to set dynamite, 1920

in Winter Quarters. Just as the paymaster, E. L. Carpenter, carried the sacks of silver and gold coins fifty yards to the company store, two outlaws in overalls and cowboy hats emerged from the crowd. One put a gun to Carpenter's head. When the clerk didn't turn over the money bags fast enough, the outlaw pistol whipped him with the butt of his gun. Firing shots overhead, the two thieves galloped off on their horses, pausing to cut the telegraph wires as they left town. Leaving a twisting getaway trail that ended in Robber's Roost, an outlaw hideout in the desert of central Utah, Butch Cassidy and Elzy Lay cleared seven thousand dollars in gold that day, in broad daylight, without much ceremony or even strategy.

In 1919, Zeph and Maud, with their four-year-old son Bill and one-year-old Bobby, moved from their home in Sunnyside to Castle Gate. Castle Gate, according to Maud, was a *coal camp*, a term, perhaps even a slight pejorative, that she applied to all the coal towns they lived in, revealing her hopes that mining would be a temporary family occupation. In the early 1900s, Castle Gate was not a sizable town and "camp" was not so far off. One main drag constituted its downtown, which was surrounded by a couple hundred cottages the Pleasant Valley Coal Company had begun to build in 1912 to attract workers. Each had a living room, two small bedrooms, a little kitchen, and a back porch. There was room on each housing plot for a tiny lawn, and the mining families would plant flowers in summers when there was sufficient rain. Some were ambitious enough to plant apple and peach trees.

Courtesy of the Western Mining and Railroad Museum

The town of Castle Gate

The Big Hall provided the community with most of its entertainment, hosting picture shows, orchestras, and dances. The meetinghouse was another gathering place, and the various religious congregations would organize the celebrations that, according to my grandfather, were enjoyed with "intensity": "From the shattering sound of dynamite strung between the sides of a small canyon and set off to greet the Fourth of July to the pomp of Easter and the exuberance of Labor Day, no occasion for celebration was ignored or underplayed."

Christmas was my grandpa's favorite holiday, during which the native-born and foreign-born of Castle Gate and the surrounding towns would come together. The mines would operate to capacity during the winter months, a result of the spike in the demand for coal, and this provided a little extra cash for gifts.

Castle Gate was a diverse community inhabited by Greeks, Italians, Austrians, Scots, various Americans of European and African descent, and a few Japanese. This town was characteristic of Carbon County, which contained more than twenty-six national and ethnic groups who left behind relics now collected in the local Western Mining and Railroad Museum in Helper. Over a thousand Italians lived in the area; they brought kitchen amenities from the Old World, such as a ravioli rolling pin, a gnocchi strainer, and a white enamel lasagna pan, all of which are now displayed in a thick glass cabinet. The golden silk threads of Teresa Stella's

Processing facility at Castle Gate, circa 1924

long-tasseled bedspread are still brilliant, woven into the purple and pink. The Greeks opened coffee shops in every mining community, providing moustache cups to prevent Greek men from dampening their facial hair. One man left behind one such cup—an ordinary porcelain mug, except for the moustache-shaped barrier with a little round opening for the coffee to flow through.

A photograph of the Tamburitza Band in Sunnyside, Utah, playing for the christening of the flag of the Croatian Fraternal Union and Slovenian National Benefit Society attests to both the diversity of the region and the focus on entertainment. Lodges were commonly founded to protect the interests of ethnic and national groups and help in their process toward achieving American citizenship. These Croatians and Slovenians performed with stringed instruments brought from Yugoslavia. In the forefront of the photograph of the Tamburitza Band, a girl with a mop of curly black hair and a poofy white dress poses holding a banjo, and a small, wide-eyed boy holds an instrument shaped like a mandolin.

Williard Craig, an inhabitant of Castle Gate from my grandfather's generation, insisted in his oral history that there was no segregation between miners of these various nationalities and races. In 1974 he claimed: "There is one thing I will have to say about a little town like Castle Gate … The majority of us are pitch engineers, retired coal miners. No matter what happens, we are just like one big family … I remember one time a person's house burned down. In less than forty-eight hours they had another house and it was all finished. That is the kind of people that coal miners usually are."

During Craig's years living in Carbon County, miners of different nationalities and ethnicities gradually assimilated, learning to deal with and even embrace their diversity. Back in the 1920s, however, coal houses were small and closely packed. Miners from that period tended to congregate with their countrymen, developing urban networks based on national loyalties. Southeastern Europeans and African Americans were brought in by companies in response to strikes, and they took over the mining jobs of white Americans and immigrants from the British Isles because they were more desperate for employment and would accept inferior salaries. Many of them likely met with a welcome similar to the treatment of the Chinese and Finns in Winter Quarters.

Even so, there were instances that justify Williard Craig's assertion of a brotherly and tolerant society. The Japanese who worked in Castle Gate were usually single men and among those most discriminated against. Still, in 1923 when the 7.9 magnitude Great Kanto Earthquake hit Japan, killing 140,000 people, the town planned a benefit concert and dance, with the profits dedicated to a Japanese earthquake relief fund. What was most remarkable about the earthquake fund is that only a year before, there had been a significant uprising in Castle Gate during the nationwide strike of 1922. Many of these pit miners from Castle Gate who scraped up cash to send to the Japanese had just returned to their homes from the strikers' tents and were pulling from the meager resources of only half a year's pay.

Maud and Zeph's family would have had an advantage in this coal camp of so many nationalities. With lineages primarily from the British Isles and their Mormon religion, they would have belonged to the majority in a town slivered into at least half a dozen significant minorities based on nativity and faith. I don't know the extent to which Zeph and Maud interacted with their diverse neighbors, only that fairness was a principle Zeph valued, and as a supervisor, he refused to privilege his co-workers of his own race and religion.

9

The Striking Years

Zeph and Maud had moved into the mining camp of Castle Gate in search of opportunity. In Sunnyside, Zeph had proved his supervisory skills as a boss driver, the man responsible for all the movement of the coal along the tracks within the mines. Maud, after years in the classroom, was an expert at encouraging potential in others. Her confidence in Zeph gave him the courage to apply for the open position of foreman of the Castle Gate mine, the top post in mining hierarchy under company management. In 1921, he was placed in charge of several hundred men—the American-born, Welsh, Scots, Greeks, Italians, and Slavs—just a year before they walked out on their jobs.

It is impossible to build an accurate picture of mining in the early twentieth century without mentioning the strikes. Welshman, for all their historical abuse and small reputation in England, were the among the first coal miners to unite for better working conditions in Britain. Before Evan and Magaret Thomas were born, Merthyr Tydfil experienced its most dramatic miners' revolt in the Merthyr Riot of 1831, where the whole town rose up against the ironmasters for rights and higher wages.

Zeph, unlike his Welsh predecessors, was not a union man, but he respected the right of workers to strike for better salaries. Union sentiment had begun to take hold in Utah before the turn of the century, and unions continued to attract members throughout the following years. A year after the Winter Quarters explosion in 1901, coal prices dropped, and Utah miners refused for the first time to take up their picks and enter the mine. By this time, Zeph was somewhere between Oregon and Alberta and no doubt felt the ripples of strike sentiment in those coalfields as well. The major strikes in Utah mining history correspond with the strikes across the United States. Coal prices varied according to the worldwide economy, much as energy prices do today. In Winter Quarters, Evan would probably have been in the forefront of the picket line with his younger sons, imposing the full force of his just over five foot frame and demanding fair wages in his hotheaded eloquence. He was a man with a clear sense of dignity and had little tolerance for mistreatment. Zeph had inherited some of this from his father, but he learned how to accomplish more with a few well-directed words than with the whole fountain of his father's fury.

After the Thomas family had moved to Cleveland, Utah, unionizers from the United Mine Workers of America (UMWA), an organization founded by British immigrants in 1890, began to infiltrate the state and recruit members, mobilizing the men for the strike of 1903–4. Strikes bore an ironic similarity to holidays: all the restless energy bottled up from working long hours underground burst out into songs and demonstrations. Miners organized parades to emphasize their demands. During the 1903–4 strike, the Italian brass band led a march in Castle Gate. Violence always simmered under the surface, and both the company guards and the miners were wary of making good on threats, lest the collective agitation of so many idle men combust and mayhem break out. The strikers hoped to outwait the company, and the company strove to outwait the strikers. Typical miner aggression involved no more than turning over vehicles and blocking the paths of coal trucks and strikebreakers. The company responded by throwing the miners out of their homes and hiring others to fill their jobs. In 1904, the tension reached such a pitch that the governor of Utah ordered the National Guard to maintain order in Carbon County.

According to J. L. Ewing, a sergeant in the Nephi Company who was sent to Castle Gate, many of his fellow militiamen went to the mining camps in support of the company and returned with reverse sympathies: "The striking miners are just as much in need of protection from the hired guards of the trust, as the company property and men are in need of protection from the striking miners ... We have but little sympathy for the miners out there, and we have still less sympathy for the coal trust. In the first place, if the trust is paying $3.00 to $8.00 a day to miners as they claim, or an average of $3.50 to $3.75 a day, they have no need to hire Finns or Dagos [a derogatory term for Italians or unassimilated people]. They can get the best miners in the country for such pay."

During this unrest, a handful of colorful labor leaders found their way to the remote mountain ranges of Utah. Charles Demolli, an Italian immigrant who had escaped from prison in Italy for counterfeiting charges, drifted among Western mines and Italian newspapers until he found himself at the head of the labor movement in Carbon County, stirring up his countrymen. Just as Salt Lake City upper crust gossip made headlines in Carbon County newspapers, the news of mining union activity was a hot topic in the city two hundred miles away. *The Salt Lake Herald* nicknamed Demolli "the silver tongued" and asserted that "his appearance is far from being the wild-eyed anarchist he is pictured by his enemies. A tall handsome man in appearance, dressed in rough chinchilla jacket, flannel shirt, corduroy trousers, and laced boots of a miner, he has a handsome face, typically Italian with a small sloping moustache."

In 1904, the UMWA brought in Mary Harris (Mother) Jones, an Irish immigrant who lost her husband and three children in a yellow fever epidemic in Tennessee. She was introduced to the conditions of the working class in Chicago, where she supported herself as a dressmaker. After her shop was destroyed in the Great Fire of 1871, she became involved in the workers' movement, helping to found the Wobblies (Industrial Workers of the World). She emerged as one of America's most powerful labor leaders when she brought the issue of child labor to the national agenda by organizing a demonstration of Pennsylvanian children who worked in the mills and mines. They marched through the streets of Washington, D.C., chanting: "We want time to play. We want to go to school." They strode right up onto the lawn of the White House, where newly elected President Roosevelt refused to step out on his porch and acknowledge them. This stunt earned her the title of the Most Dangerous Woman in America.

An ample, kindly-looking woman of seventy-four, she chose to announce her presence in Carbon County in characteristic style by descending on the best hotel in Helper. After her meal, when the waitress brought out a finger bowl for her to wash her hands, Mother Jones proclaimed: "Take it away my girl. Such things are not for me, they only give some poor overworked girl extra work at washing dishes." This denunciation, surely meant for the wealthier community members who dined around her, was passed from mouth to mouth among the miners all through Carbon County and deemed a noble declaration. Gossip of Demolli and Mother Jones would have reached the Thomas home in Cleveland, and I wonder what my great-great-grandmother Margaret thought of her times—if she was caught up in the crusading spirit of the fight for workers' rights, or if she watched with wonder and uncertainty the conflict between the company and the workers, led by these larger than life personalities.

From the Helper hotel, Mother Jones moved to a tent colony, rooming with an Italian family among the strikers, keeping up their spirits. In the dry lands of central Utah, conditions in the strikers' colonies were terrible. Average precipitation in the region is between five and ten inches per year, which would have forced residents to haul water from deep wells and springs. The Utah Fuel Company had a near monopoly on the stores in the area, and the mining towns were, for the most part, far from farming communities. The mountains that provided a suitable climate for brush and pine would have supported little provender for foraging. Deer and jackrabbits would have been the most available food source, aside from the company's canned and bagged goods.

Such a hotbed of squalor made local law enforcement nervous. The arrival of Mother Jones only promised to heighten the strikers' determination, so the sheriff of Carbon County organized a posse of forty-five men to raid the strikers' colony and drive the troublemaker out of town. Mother Jones wrote in her journal: "Between 4:30 and 5 o'clock in the morning ... They descended upon the sleeping tent colony, dragged the miners out of their beds [without time] to put on their clothes. Shaking with cold, followed by the shrieks and wails of their wives and children, beaten along the road by guns, they were driven like cattle."

Mother Jones had gotten word of the raid and convinced the Italian miners to bury their guns to prevent any bloodshed. As a consequence, 120 Italian miners were forcibly taken from their lodgings. The miners made no resistance, and contrary to Mother Jones's account, the men who had been pulled from their beds in their nightclothes were allowed to return to their tents and change. Their families were driven from their tents, and the women and children shivered in the night while soldiers searched their homes for munitions. The men were loaded into a boxcar headed to Price City, their wives and children running and screaming after the departing train. The Price City jail could not hold so many men. Consequently, company employees erected a makeshift bullpen around a warehouse used to store carriages, herded in the Italians, and put up armed guards to keep them there. Meanwhile, Mother Jones remained behind to comfort the women. Two days after the raid, she reported: "The stone that held my door was suddenly pushed in. A fellow jumped into the room, stuck a gun under my jaw and told me to tell him where he could get $3,000 of the miner's money or he would blow out my brains."

Mother Jones handed over the fifty cents in her pockets, saving her brains and little enriching this "fellow," who turned out to be Gunplay Maxwell, a pardoned bank robber who had been sworn in as a deputy. He was the bodyguard of company lawyer Mark Braffet.

A photograph from the time shows both Maxwell and Braffet standing watch over the bullpen of Italians, who hang their arms at their sides. The prisoners gaze directly at the camera. Their faces in the photograph are too small to discern whether their eyes are glaring—ordered to attention by the men stationed with rifles on the roof above—or if they are posing, aware even in their extremity of the need to capture such a historical moment. Many union members were held here for trial, some detained as long as a month. Local hotels and residents were asked to cook meat for the prisoners, and the local Mormon bishop gathered food from among his parishioners. Italian women would wander to the site, carrying food and pleading with the guards in broken English to free their men.

Zeph would have been familiar with these stories, reading them in the newspapers and hearing them by word of mouth from his coworkers. Certainly there were similar episodes in Cleveland. Likely members of Zeph's immediate family participated, cajoling him to throw his support behind the union. Zeph always declined because he valued order and refused to negotiate by means of violence.

In the years that followed, strike activity continued, but it was poorly organized due to competition among the unions. In 1914 World War I started, increasing the average wage of the miners in tandem with the increasing demand for coal. After the war, the expansion of the mining industry glutted the market, and both the demand for and price of coal plummeted. It was in these troubled times that Zeph took up the position of foreman at the Castle Gate mine. In April 1922, over twenty thousand coal miners prepared for the largest strike in American history up to that time.

I can only imagine the interchanges between Zeph and the company in the early days of that month. He spent his time among the men and was likely familiar with the labor agitators who moved among them. His living conditions were only slightly better than those of his fellow miners, and he knew what a pay cut would mean to their families. Zeph's job was to run the mine and supervise the workers, but the company was solely concerned with making a profit. Despite the threats of the miners, the Utah Fuel Company reduced wages from $7.25 to $5.25 per day. Zeph would have been the harbinger of this bad news. As a result, in solidarity with the coal miners across America, 70 percent of Utah miners walked out.

The Utah Fuel Company responded with intractability. The striking miners were evicted from their houses, although it was early spring and there was still snow on the ground. They were given five to ten days' notice, and the belongings of the miners who refused to leave were tossed out of their homes. In Sunnyside, as company officials attempted to dispossess a woman in labor, the company doctor stood on her porch, threatening to shoot any who dared. Reports in newspapers villainized the foreign-born, who were represented as being the agitators of the turmoil, and racism was the momentum behind the recruitment of strikebreakers. In order to counter this negative press, Hideo Kazuta, a Japanese immigrant who served in World War I, wrote the following to one of the local newspapers: "I love America. I love the state of Utah and especially Carbon County … I stand against lower wages in order to save Kenilworth where I was working … from utter misery, and to protect the businessmen and farmers of Carbon Counties where the workers' wages are the barometer of prosperity. Every genuine American I feel is opposed to the present tendency to

Zeph Thomas (second from left) posing with his coworkers in Castle Gate, circa 1924

change our standard of living in our beloved country by lowering wages to that of other lands where kings and a few followers have the lion's share of the nation's wealth and happiness."

Striking miners formed tent colonies in the severe early spring conditions, supplied with materials that had been shipped from Wyoming by the UMWA. Squatting on tracts of lands just outside the mining camps, miners would do what they could to interfere with mining operations. The company officials responded by erecting wire fences to close off the communities. During the workday, children would entertain themselves by climbing the fences to throw rocks at the strikebreakers. Small shootouts among the adults were common, with a handful of men wounded on both sides. In Scofield, teachers would keep children in the schoolhouse after their lessons, instructing them to lie on the floor to avoid stray bullets. The first man killed in 1922 was John Htenakis, a Greek miner, who was shot by Deputy Sheriff Lorenzo H. Young.

This event polarized the conflict. Young claimed he shot Htenakis in the chest in self defense, a statement that was supported by the company doctor. The Helper doctor sided with the Greeks, who maintained Htenakis was shot in the back. At his funeral, the Price band played as his casket was trailed by seven hundred Greeks, carrying their blue and white national banner.

Not long afterward, the Utah Fuel Company brought in a train car full of strikebreakers from Colorado, and miners set up an ambush around

Bill, Bobby, and Boyd, circa 1922

the tunnel between Castle Gate and the neighboring mining town of Standardsville. Company guard Arthur Webb had volunteered to shovel coal into the engine to keep the train going. When the train passed by the ambush site, the strikebreakers and the miners exchanged fire; one miner was hit in the shoulder, and Webb was killed. The death of Arthur Webb settled the score of John Htenakis as far as the miners were concerned, but it prompted the governor of Utah to call in the National Guard and establish martial law. The miners' guns were confiscated, and machine guns were positioned above the strikers' tent villages. At ten o'clock, the miners were ordered to extinguish all their lamps, and the darkness of the canyons of central Utah was cut only by searchlights sweeping over the tent cities through the night.

My grandfather Robert remembers these times vividly. He was a boy of four, but the general air of agitation terrorized his early memories. In his personal history he writes: "One does not easily forget the probing searchlight, which turned my bedroom into garish day many times at night during the strikes at the mine. My father told us honestly (he answered all our questions honestly) that the company was afraid someone might set fire to our house, but that we really had nothing to worry about because the strikers were our friends and neighbors."

As the foreman of the mine, Zeph was placed in the precarious position of being the company's representative to the miners while he felt sympathetic to the workers' cause. The company issued him a .38-caliber pistol for protection and insisted he'd need it.

Zeph laughed and said, "I think I had better go down and talk to the miners."

"You can't step into that tent city. They'll kill you," they countered.

Zeph drove down from Castle Gate two miles to the strikers' camp, bringing his boys along for the ride. He walked out among the tents and talked to the people. Self-effacing and a patient listener, Zeph was a quiet man. He had worked his way up from the bottom, and the miners knew he was one of them. Although he had to maintain his loyalty to the company and had little sway over wages, he listened to the workers' complaints. From inside the car, his sons likely watched the children of the striking miners chasing each other, making mischief out of boredom and rumbling bellies. In the end, the miners held their ground for five months, without work or pay, before the Utah Fuel Company gave in to their demands.

Bill was seven and Bobby was four. I don't know if they stepped out of the car with their father. There were few amenities in the tent colonies. The human smell of bad sanitation, the noise, and the agitation must have terrified them. Zeph wouldn't have brought the two boys just to spite the company, and even less to ensure his own safety; there must have been something he wanted his sons to understand. I suspect it might have been the fact that Zeph knew they were not destined for lives bound by coal—the family was not in the same desperate state financially as the striking miners. Maud had been a schoolteacher. The boys were bright and eager learners, and their mother kept their minds active. Bill became a chemist, Bobby a scholar, and Boyd, the youngest brother, a judge, but none forgot the mining camps of their early years.

10

Get the Men Out

Before Zeph had applied for the job of foreman at the Castle Gate mine, he was required to pass the state licensing exam, a written test that would have been a significant challenge for a man with a sixth-grade education. The job meant a significant raise in salary and a promotion in responsibility; most foremen supervised an average of a hundred men. In 1918, the US Bureau of Mines listed the following job qualifications:

> Physical strength; good health; more than average ability; ability to handle men. He should have large practical experience and a general knowledge of coal-mining conditions as to methods, safety requirements, etc. He should have had experience as pusher, repairman and coalminer, or engineer, and be capable of reading mine maps and laying out work at coal mines. He should have a certificate of competency.

Maud, although never adept at sewing, cooking, or homemaking—she was always off reading when her mother tried to teach her these womanly arts—was confident in the tasks of the mind. She began tutoring Zeph, drilling him on the questions in the practice exam. This went smoothly at first, but then they reached an impasse.

"What is the first thing you would do, and what procedure do you follow, if the electricity stops and the fans go out?"

"Get the men out," Zeph replied.

Maud must have teased her husband encouragingly, "Now wait a minute. You have a whole page, and all you said was, 'Get the men out.' Come on. What do you do first? Then finally, of course you get the men out."

But Zeph had learned one thing from his father—a quality in Evan that was probably called obstinacy, but in Zeph took on the cast of conviction. "No, no. You don't do anything else. You get the men out." No matter how many times Maud quizzed him, his answer remained the same. She finally gave up and hoped that the question wouldn't appear on the real exam.

But this question came up first, and Zeph wrote, doubtless in a cursive markedly less elegant than his wife's, "Get the men out."

The general inspector stopped him later that day to tell him, "Zeph, you created a bit of a sensation with your exam. You were the only one who said what we wanted everyone to do. Get the men out. We left that big blank so that people would fool around with it. But Zeph, you were the only one who said, 'Get the men out.'"

Although he respected the wisdom of his wife, this was an answer Zeph knew from his experience in the mines, a world that Maud saw from only the outside. An inquisition into the explosion at Winter Quarters revealed that had the men run into the corridor leading to Mine No. 1 instead of Mine No. 4, their lives would have been spared. Evan Jr. and Frederick had died from gas, not from the explosion. Had the miners been instructed better, Zeph wouldn't have lost his older brothers.

Zeph was well aware that as foreman, in addition to making hiring decisions, assigning rooms to the tonnage miners, and managing other daily activities, he was primarily responsible for the safety of his miners.

Among the photographs we have of the mining camps is one from 1924 after Zeph had been a foreman at Castle Gate for three years. Zeph, in a cap and tie, watches his crew compete in rescue exercises. Five young men angle back in a line, all wearing round helmets with fishbowl-like glass over the opening in the front. Underneath, near the mouth, two rubber tubes like walrus tusks attach the helmet to the oxygen tanks that the men carry on their chests. The man in the foreground has rolled-up white sleeves, revealing a young, wiry arm. His overalls are cuffed twice. He turns his helmet toward the camera, peering through one bulbous eye. "Co. ANNUAL FIELD DAY" is printed on a white banner strung along a brick building behind. The rescue equipment Zeph's crew wears is a similar model to those the rescue crew wore in retrieving the bodies from Winter Quarters.

Once appointed foreman, Zeph refused to hire or promote the workers he supervised based on race or nationality, a fact that put him at odds with some of his Welsh friends and relatives, but earned him wide respect for fairness. The years following the 1922 strike continued to be bad ones for coal. Castle Gate No. 1 was closed due to a lack of orders for coal, and in March 1924, Zeph assigned all married miners to Castle Gate No. 2 to ensure that they would have wages to feed their families.

Castle Gate was a mine known for taking precautions. In 1892, a territorial inspector had visited and come away favorably impressed, claiming "that altogether this is a mine where no expense is spared to provide for the safety of its workers and methods that are in daily use here, which are not found in any other part of the mining world." Castle Gate

Zeph's crew competing at Annual Field Day, 1924

had a sprinkler system to keep coal dust down even before it was discovered to be flammable because it had been known for many years that coal dust caused black lung. To prevent deaths from cave-ins, shots were fired electrically when all miners were out of the mine. In 1924, Castle Gate received two more technological improvements: a machine that had been developed to undercut coal from the bottom and electric headlamps to replace the open-flame carbide lamps. On March 8, these electric lamps were still packed in boxes, waiting in the storehouse for the charging apparatus that had not yet arrived.

For miners who had only been working one to two days a week, the eight-hour shift offered on March 8 in Mine No. 2 was considered a boon. Many residents of Castle Gate, however, felt uneasy about going to work that day. Fay Thacker reported having a strong feeling that he shouldn't go in Mine No. 2. He went to the engineer who was making improvements in Mine No. 3 and asked for a job for himself and three of his friends. Otto MacDonald, a farmer to whom Fay and his wife gave shelter during the winter months, decided to wait a few days before returning to his farm and took advantage of the eight-hour shift offered in Mine No. 2. Benjamin Thomas was a butcher and the LDS bishop of Castle Gate. Throughout his life, he had commented to his family and friends about having been given strong spiritual warnings not to work in the mine. His shop closed when the town fell on hard times, so he worked at the tipple above ground. On March 8, the eight-hour shift was too enticing, so he decided to ignore his repeated premonitions and asked for a slip to go down. For no apparent reason, mine plumber Frank Mangone decided not to go to work. His wife asked him if he was sick. He replied, "I'm not sick. I'm just not going," and spent the day with his children, who were thrilled to have their father home on a Saturday.

At six thirty in the morning the miners took the mantrip to their assigned rooms. The fireboss made his regular rounds in the mine with a miner's safety lamp, the flame of which was encased in a cylinder of tightly woven mesh to prevent it from igniting the gases outside. When a flammable gas was present, a blue halo would form around the tip of the flame in the lamp, and the length of the halo determined if the levels were dangerous and the mine needed to be evacuated. That morning the fireboss found a pocket of gas in room number two, near the roof where coal had been shot down the night before. He climbed on top of the pile of blasted-out coal to investigate. His headlamp was blown out, so he stepped back down to relight his lamp from the flames of two workers loading the coal.

At the mouth of the mine, Zeph was called to the phone just as he was about to escort his good friend, mining inspector Jack Thorpe, into the tunnels. Zeph excused himself, and Thorpe entered the mine without him. On the telephone, Zeph heard only the first words of the local blacksmith, who called to talk about the repairs on Bill's bike, when he heard a boom and the mountain shook. He ran to the entrance where, according to one man's diary: "smoke and gas was coming from all three entries so thick that if a man walked up to the mouth of any … and stood there for two or three minutes he would have been overcome with gas."

Later, it was determined that the fireboss ignited the methane from the roof when he relit his lamp. The initial blast split as it exited the room, with the main force blowing out toward the open air where Thorpe had just stepped inside the mine. His wife Eva had dreamt the night before about "a ball of fire that shot many streamers of fire to all her neighbors. When the last streamer of fire hit their house she screamed, 'Jack you're going to burn.'" A secondary gust rushed toward the interior of the mine, where it lost much of its force and merely blew out the lamps of the men working in the farthermost section. The air was full of coal dust from the explosion, as well as the dust that had accumulated during the days when the mine had lain idle. (According to tests by the US Bureau of Mines, Castle Gate produced the most flammable coal in the country.) When the miners in this section relit their carbide lamps, a second explosion ripped through the rooms just seconds after the first,

Courtesy of Western Mining and Railroad Museum

Townspeople waiting for rescuers after Castle Gate explosion, 1924

killing the remainder of the 171 men in the mine. Twenty minutes later, a third boom caved in the main portal, blocking the first thirty feet. Not one of the miners escaped alive. Jack Thorpe was killed instantly; Zeph was saved only by a blacksmith's phone call.

After the sound of the explosion, a whistle shrieked out into the town to indicate there had been an accident. A man working in the mining office across from the Thomas's home ran up to the door and told Maud, "Zeph's okay." Bobby stepped out on the porch. He was five years old and didn't entirely understand what had happened, but he remembered more clearly than any other image of his early years the women running past his house to the mine, crying and screaming.

Rescue efforts started immediately. Engines from the Price and Helper fire stations sprayed water into the mine to put out fires, but their efforts were useless. Rescue teams put on forty-pound packs containing oxygen and breathing apparatuses—goggles had replaced the astronautlike helmet Zeph had worn for mine-safety training. From the blast of the explosion, eight-inch thick water pipes were "turned out like straws." The sprinkler system, which was supposed to prevent coal dust from igniting, spewed into the mine from the twisted pipes, filling some parts with up to ten feet of water. Thomas Hilton and three other men entered the mine first to turn off the water to prevent any injured men from drowning. They carried in a rope to serve as a lifeline. One hundred yards in and halfway to the valve, Hilton began to feel dizzy and ran for the outside. Near the entrance, his legs buckled and he fell unconscious. Zeph and the general manager, Mr. Littlejohn, put handkerchiefs over their faces and dragged Hilton out. They carried him to the Castle Gate cemetery, not far from the entrance, and propped him up on a tombstone until he regained consciousness.

Rescue crews arrived from all over Carbon County. Even a coal mine in Wyoming volunteered to send in a rescue crew. Miners from Peerless, Winter Quarters, Standardsville, Sunnyside, and Castle Gate were divided into three shifts of eight crews each, which entered the mine on a rotating basis as the previous group's oxygen supplies were spent. Salt Lake residents and pet store owners donated forty-three canaries to help detect the presence of afterdamp, the carbon monoxide gas that had killed 123 miners at Winter Quarters. The rescuers would descend down the corridors of the mines where fires still burned and rocks and debris obstructed their path. In front of their masks, they would hold a canary cage, and once the bird fell from its roost, the miners knew they were in danger and would not press on. The public, reading the news of the disaster, were assured that the birds would "recover under treatment."

One of the rescuers, George Wilson, captain of the Standardsville team and a seasoned miner, was killed in the rescue efforts when the mask of one of his crewmen malfunctioned. Wilson carried the man to safety, but the miner's flailing arms knocked off Wilson's nose clip. Five minutes later, Wilson died from inhaling the poisonous gas, adding another mining casualty to a family who had lost three sons and a father in the Winter Quarters explosion.

Rescue efforts persisted around the clock, and at one in the morning on March 9, the first body was carried out of the mine. By the next day, twenty-six dead men had been retrieved. Because they had all been killed by the explosion instead of gas, many were burned and mutilated beyond recognition. A local newspaper described the scene: "As sweating miners with lamp-lit caps emerged from the tunnel with their comrades' bodies, up above them on the road, a hundred yards away, this morning stood for a time a group of wailing women, wives of alien [foreign] miners who lifted their appeals for mercy to the heavens. 'O dolor mia, o dolor mia,' cried Italians, while Austrians and Serbians wept out their grief in their own tongues. It became necessary to send them home; their unrestrained sorrow was too contagious."

Zeph (far left) with Castle Gate rescue crew, 1924

Store owner Aton Dupin, a Croatian immigrant who spoke English with a southern accent, had worked in the Castle Gate mine prior to opening his shop. He volunteered to load the bodies into his delivery truck, driving them to the amusement hall, where they were laid out to be identified. After performing this service, he fell ill for the first time in his life, recognizing his comrades among the dismembered corpses. The women were advised not to ask to see the bodies of their husbands, sons, and fathers, but to remember instead how they saw them last. Some still insisted. A newspaper reported: "All day Tuesday and Wednesday hundreds of people visited the morgue. In the amusement hall the dance floor … is filled with caskets of various shades and colors. Many of the women have become hysterical and have had to be led away from the caskets of husbands and sons as they have been permitted in some instances to look at the faces of their loved ones. After Wednesday, the company doctor forbade any further viewing of the dead beyond purposes of identification to protect the sanity of the living."

Maud did not see Zeph for many days. Occupied with coordinating rescue efforts and identifying the dead, he did not return home until after March 18, when all 171 bodies of the miners were removed from the mine, totaling 172 dead with the addition of George Wilson. The women whose men were not killed in the explosion joined forces with the Red Cross to cook meals for the rescue workers and the grieving families.

The final tally of the dead listed seventy-six American-born (including two African Americans listed as "Negros"), forty-nine Greeks, twenty-two Italians, eight Japanese, seven English, two Scots, and one Belgian. Of

Courtesy of Western Mining and Railroad Museum

Caskets in Castle Gate Community Hall, 1924

those, 114 miners were married, and they left behind 417 dependents. Nobody blamed my great-grandfather for his decision to give the work to the married men. Regardless, Zeph felt responsible and volunteered to resign his position, a decision supported by Maud, who wanted to leave the dangers of the coalfield. When reporting his decision to his superiors, he committed to getting the mine back in working order, which was at least a year's project—the coal-cutting machines had been ripped apart by the blast; an eight-hundred-pound steel door had been blown a half mile across the canyon; roof timbers and rails sailed as high as five thousand feet; and the coal along the tunnels had been coked by the extreme temperatures.

Religious and secular leaders from the surrounding communities flocked to attend the funerals. They included Utah Governor Charles Mabey, Archbishop Peter Fumason-Bondi (an apostolic delegate from Washington State), Bishop Joseph S. Glass of Salt Lake City, Italian Consul Fortunato Anselmo, and President Heber J. Grant, the prophet, or head, of the LDS Church. President Grant had traveled from Salt Lake City, and he stayed with Zeph and Maud. For the Mormons of this small town, a visit by the prophet would be regarded as a significant event. He was a leader as respected in the LDS ranks as the Pope is among Catholics. My grandfather remembers him, a man of small stature, standing in the living room with tears running down his face. He had been out among the women and knew nothing he could say would comfort them.

Since the Winter Quarters explosion, workmen's compensation laws had been put into place, which included state-provided death insurance. The Utah Fuel Company was independently insured, and it covered the funeral costs, in addition to sixteen-dollar weekly stipends to widows for a period of six years. Governor Charles Mabey designated a committee to distribute the $132,445 that had been raised by public donations. After the disaster, life was particularly hard for the Greek, Italian, and African American widows, who were not assimilated into general society. According to Greek culture, women's and men's spheres were separate, with many Greek women not venturing far beyond the home. By custom, widows did not remarry, and several of those from Castle Gate used the money given to them to return to their native country. Most of the Italian widows who stayed in Carbon County took new husbands; the two who remained single ran a bootlegging business to survive. Making ends meet was even more of a struggle for one African American widow who was not legally married and was granted compensation funds only after the issuance of a court order.

According to one historian: "With immigrants from the Balkans and Mediterranean initial discrimination was strong but of relatively short duration and never so violent as that experienced by Blacks. Public opinion kept these new immigrants out of certain areas of towns and cities and frowned on intermarriage." The African American miners who had entered Utah as strikebreakers in 1922 were discriminated against not only for their color, but also for taking over other miners' jobs. One exception was Prince Alexander, who had come to Carbon County in 1914 and was among those killed in the Castle Gate explosion. He was an educated man, formerly a preacher, and had impeccable manners, which initially drew the mockery of his fellow miners. But in the flu pandemic of 1919, Prince nursed the sick in Castle Gate, gaining a reputation as a savior to many in the camp. After the March 8 disaster, he was buried in the Castle Gate cemetery next to miners of other ethnicities. The two men listed as Negroes in the newspaper's enumeration of the dead were Ed Willis and Archie Henderson; In death, Prince Alexander had become assimilated into the general category of American. His coworkers might have considered this an honor that Prince had somehow transcended his race. Although the racism inherent in such a sentiment is evident, Prince was a man who was able to look beyond the prejudices of others and help them look beyond their own. Likely, my great-grandfather Zeph knew him, and I hope he admired Prince, as I do.

11

Leaving Carbon

Six months after the explosion in Castle Gate, the unity that had been developed by the collective tragedy was ripped apart. Strangers drove into the neighboring town of Helper and erected a cross on the hill. This was the first of the burning crosses in Carbon County, where the Ku Klux Klan began to recruit members. The Greeks, the Italians, and the Slavs in these communities formed spy networks to unhood Klan members and erected flaming circles on the hills in opposition to the crosses. My grandfather Robert was six at the time, and he mentioned these in the troubled recollections of his childhood: "I can't remember precisely how old I was when I saw my first fiery cross burning a few hundred yards up the mountain to the rear of our house. I could barely discern the hooded Klan figures near it, but from the despairing look on my father's face, as he too, watched, I sensed all the ugliness, bigotry, and hatred that the flaming cross carried."

An Italian girl of about the same age, who is remembered in the historical record as Joe Barboglio's daughter, interpreted the burning crosses similarly. However, to her they had a more personal application. She knew they meant that "there were people who hated us, and all the other foreign-born families." That she is noted as "Joe Barboglio's daughter" is significant. After the rough treatment of Italians in the 1903–4 strike, Joe Barboglio left Castle Gate and settled his family in Helper. He became a noteworthy member of that community and helped found the Helper State Bank.

In 1925, racial hatred hit a climax. Milton Burns, a company official, was allegedly killed by Robert Marshall, an African American miner. The June 15 incident is still a divisive issue in Carbon County, so even current historians retell the story with different slants, some emphasizing the culpability of Marshall and others focusing on the violent reaction of the community.

Milton Burns was the special agent for the Utah Fuel Company, assigned to keep the peace and enforce company policy. He was a native of the area, a former sheriff, the father of six children, and a man who was immensely popular with the locals. The children especially loved him because he would give them rides on his large gray horse as he made the rounds. It was commonly known, however, that Burns was a Klan member. Prior to the shooting, Marshall had had an argument with Burns and another company official and had "drawn his time" (quit and received his paycheck). Later that day, Marshall hid under a wagon bridge and confronted Burns as he was making his rounds of the company site. According to two small boys, the sole eyewitnesses to the case, when Burns tried to take Marshall's rifle from him, Marshall shot him five times, stomped him in the face, and fled. Burns died the following evening and, according to some accounts, identified his killer. A forty-man posse was sent out after Marshall, who was discovered three days later in the shack he shared with another black miner. His housemate had turned him in for the $250 reward.

According to the 1925 Price, Utah, *Sun Advocate*, Marshall was considered a "bad actor," having already put bullets into both a girl in Rolapp and another black man named Ras Jackson in Grand County. After apprehending Marshall, company officials drove him down to the Price Courthouse. There an angry mob of local citizens seized Marshall from the car and drove him, in a caravan one hundred cars long, three miles to "the hanging tree east of Price." They strung Marshall up, lifting him thirty-five feet into the air, where he dangled for almost ten minutes until deputies arrived to cut him down. After a few minutes he revived, and someone suggested that he be shot, but the crowd yelled, "No! Let him suffer." The deputies tried to put Marshall back into the police car, but he was wrested from their hold once more and hung again, this time being jerked around until he died from a broken neck.

A reporter from the *Sun Advocate* indicated that the crowd wasn't the typical one you'd expect for a group of lynchers: "Your neighbors, your friends, the tradespeople with whom you are wont to barter day by day, public employees, folks prominent in church and social circles, and your real conception of a 'mob' might have undergone a radical turnover ... No attempt at concealment was made by any member of the lynching party ... [there was] quite a sprinkling of women—the wives and mothers of the good folks of the town. And, too, there were even some children."

One man bragged about attending the lynching claiming, "It's a proud day for me that I helped pull the rope." Someone took pictures of the hanging and the crowd gathered around; these were reprinted as a package of postcards and sold door to door for twenty-five cents. Under pressure from the National Association for the Advancement of Colored People, Governor George Dern issued eleven warrants for the arrest of some of the most prominent citizens in Castle Gate for first degree murder. At the trial, over one hundred witnesses refused to identify them as the perpetrators, and all were released without charge.

It wasn't until 1998 on what locals called "a day of reconciliation" that Marshall was given a funeral service and a proper burial. In 1925, the African American residents of Carbon County, numbering just over a hundred, had taken up a collection to bury Marshall, but they lacked the funds to put up a gravestone. The plaque erected in 1998 reads: "Robert Marshall: Lynched June 18, 1925, A Victim of Intolerance. May God forgive." Marshall was the last man to be lynched in the American West.

Zeph knew many of the men who had been involved in the lynching, and he was fundamentally disturbed by the collective air of approval surrounding what local residents considered to be an act of "justice." Zeph had consoled the miners during the strike of 1922 and wept over their corpses in the explosion of 1924. As a foreman he had always exhibited fairness and compassion, leading his workers by example. The lynching of Marshall signified that his efforts of tolerance had not tempered the rash and rough character of the mining town he had served for four years. Since the explosion, Maud had been pressuring Zeph to leave—she couldn't bear the thought of him dying in the mine—and now Zeph had his own reasons to move on. In the fall of 1925, Zeph and Maud packed up their belongings. With their three small boys, they drove past the two monumental walls of the Castle Gate to Provo, never to return to Carbon County again.

In Provo, they bought a small farm (part of Maud's scheme), but Zeph could never make it profitable. He applied at Geneva Steel, a company where the coke from Castle Gate was used to fire the steel furnaces, but he was unable to get work because his skills were so specialized for coal mining. Maud and Zeph took in boarders, both of them doing the cooking and washing dishes. A year later, Zeph was offered a job developing a small mine in Coos Bay, Oregon. Maud cried over the prospect of leaving Provo, but she decided to support her husband in the work he knew best. Zeph assured her that his return to mining would only be temporary.

In Oregon the mines were more primitive. Everything was modernized and electric in the Castle Gate mine, but Zeph's Oregon mine was run primarily by manual labor. Located fourteen miles outside of the nearest town, it could only be reached by an unimproved road that disappeared in the rainy season. The coal was of lower quality and had to be carried out of the mines by mules.

Zeph loved mules almost as much as he loved people. Sensing this, his mules would follow him around like dogs, waiting to be patted or rubbing up against him. But Zeph was no pushover. Mules can be difficult to

Miner and mule in Coos Bay, Oregon mine

manage, and if they misbehaved, Zeph would whack them with something that made a loud noise or twist a rope around their noses to get their attention. My grandfather Bob remembered a particular mule named Barney, who was small but smart and often was the lead mule pulling the cars out of the mine. One day, Barney was groaning and jerking, struggling to start the car of ten tons of coal. The boys yelled, "Come on, Barney," trying to encourage the little mule. Zeph walked up and hit him.

My grandpa reprimanded his father, saying, "Oh, Dad, that poor thing's trying."

Zeph stuck his hand between Barney's neck and the harness to show that the mule wasn't even touching it. Knowing the jig was up, Barney leaned up against the collar and pulled the coal out. My grandfather observed that the mules never resented Zeph for his discipline because he was fair. Although this assertion is hard to justify, Zeph did ensure that the men who worked for him treated the mules humanely. One man who worked for Zeph always pestered the mules and jabbed them in the ribs. One day he was throwing lumps of coal at a mule because it stopped pulling a car. The animal had intentionally halted in the narrowest passageway of the mine, waiting for the miner. When the man sidled by to yank the mule forward, it leaned against him, pressing the man's body against the wall of

Bob, Bill, Maud, Zeph, and Boyd in family picture, Coos Bay, Oregon, circa 1936

the mine with all of its 1,200 pounds. Zeph heard the man's screams, and when he walked into the room, the mule let up. The man slumped to the ground. Zeph assessed the situation and reprimanded the miner instead of the mule: "Don't nag him every time you go by. If there's something he needs to do, you see that he does it."

Zeph's mine was never very successful; during the Depression, demands for coal were low, and the family was much more financially strained than they had been in Castle Gate. Because they always planned to return to Provo, Zeph built only a simple cabin for the family in Oregon, and all their fine furnishings remained in storage in Utah. Once it became apparent that the move to Coos Bay was permanent, Maud gave all her nice things to her sister. She summoned her energies and made the most of their situation. She helped organize a branch of the LDS Church in Coos Bay and taught Sunday school in her cabin. She started a book group among the people in the town, and they read the classics together. Most were poorly educated and unfamiliar with books, but many gained a respect for fine literature as Maud dramatized it out loud.

The boys spent most of their time outside of school working with their father. After the coal was mined and carried out by mules, Zeph and his sons would take the coal into town in a dump truck that had no brakes. Usually there was little traffic in the fourteen miles between the mine and Coos Bay, but once in a while the police would stop cars in the middle of the road. On one such occasion, Zeph had to drag the truck along the guard rail to slow it down, only stopping when he ran into the back of somebody's bumper. The truck was not dependable in other ways, and it would occasionally dump its load in the middle of town. The boys, prepared with shovels, would jump out, jam pieces of wood under the tires to stop the truck, and shovel the coal back in—a humiliating experience because the local teenagers attended the same school.

Bill, Bobby, and Boyd also helped their mother at home with the sweeping and heavy cleaning, as she had a bad heart. Under her encouragement, they maintained good grades, finishing high school with honors. Bob graduated in 1936, but he didn't have enough money for college, so he worked in a woodshop for the next three years. The Thomas family held a council and decided that Boyd would work to support Bill and Bob while they were getting an education. Both sons enjoyed a year at the University of Oregon, but they returned when Maud died of kidney failure in 1939.

By this time, Zeph had developed bronchial asthma from working in the mines for thirty-eight years, and Bob determined the only way to keep him alive was to relocate to a higher and drier climate. Bob dropped out of school and moved to St. George, Utah, where he set up a cabinetmaking shop to support his father. Despite the change to a less humid environment, Zeph's spirits never improved. His illness prevented him from working, and his idleness left him despondent. He understood the sacrifice Bob was making to care for him, and he missed Maud.

After a year, Bob closed the shop. Wood was scarce due to the war effort. Bob moved back to Oregon and began working in the shipyards, saving money for college. His father joined him there later until Zeph passed away in 1943 at the age of fifty.

Zeph's sons shipped his body back to St. George, where Maud was buried, and told their relatives that their father had died of a broken heart. According to my father, "Zeph and Maud were always more successful as a combination: Maud couldn't sew, so Zeph would read a pattern and cut the fabric, and she would sew a straight line where he directed her." Despite her bad health, Maud's vitality is what brought energy to a household supported by rigorous and oftentimes discouraging labor.

A crowd of rough looking men attended Zeph's funeral, and among them were those who had been at odds with Zeph during his lifetime. One estranged brother whom Zeph had not heard from in thirty years traveled from California to pay his respects to the brother he claimed had "terrorized" his life: "I would be having a really great time, and here [Zeph] would dance by and say, 'Watch it!' He was acutely aware of anything even bordering on impropriety ... However, he didn't expect of me what he demanded of himself." A miner whom Zeph had fired from the Castle Gate mine showed up, telling Bill, Bob, and Boyd that Zeph was the finest man he had ever known. That particular miner had had a reputation for being reckless. In charge of pulling pillars (columns left behind in the room and pillar method of mining), he had been more concerned about

Zeph and Maud, circa 1933

making money than with safety. After he removed pillars that Zeph had instructed to leave alone, Zeph sent him on his way, because the man was endangering not only his own life, but also the lives of others. The miner wanted Zeph's boys to know that their father "had been kind to him at a time when he didn't deserve kindness."

My grandpa was in charge of delivering the last address over his father's grave. He must have thought about his mother, Maud, and the many speeches she had been asked to give over the years. Bob had barely a year of college under his belt and was only twenty-five years old, but he had inherited his mother's love of literature. I imagine Bob's comments bore the mark of Maud's style: "When we say that 'Dad was quiet' we have said nothing, yet everything. No pretense or hypocrisy rested easy in his company, for he was without guile. True to himself, he was false to no man. He accepted no compromise with truth, yet had acquired a quiet tolerance which made his respect a thing to be cherished. 'Praise from the Emperor,' Mother called it—even as children Dad's few words of praise were long remembered. He expected everyone to do his best, and those who worked and lived with him tried just a little harder in his presence."

12

Ghost Towns

After burying Zeph in St. George, Bob returned to the shipyards in Oregon. He supported himself through a bachelor's degree at Reed College, graduating in 1947. Although grateful for the opportunity to study, Bob must have felt very alone. His brothers were in a similar situation, struggling their way through university degrees. Without the benefit of counsel from his parents and at a time when graduate study was far from the norm, Bob decided to pursue a master's degree in English at the University of Oregon in Eugene.

At age thirty-one, Bob was balding and had little dating experience. While teaching a Sunday school lesson at the local LDS congregation, he met my grandmother Shirley, who was among the members in attendance. When he brought up philosophy in his discussion of religion, Shirley raised her hand and asked him, "What about Hegel?"

He pointed to her and said, "I'll see you after class."

My grandmother's mom, Eva Nancy Angeline, was a southern belle and had instructed her daughter well in the art of flirtation. Seven years younger than Bob, Shirley Wilkes was petite, pretty, and well mannered, but the thing that interested Bob most was the characteristic that had attracted Zeph to Maud: her mind. They married on December 24, 1947 in St. George. Two years later, they moved across the country so Bob could pursue a Ph.D. at Colombia University. Shirley found a job teaching in a private school.

A generation removed from the mines, my father Ryan's earliest memories are of streetlights, narrow alleys between tall buildings, and a New York City bus that almost took off his arm when he was three—a very different landscape than the mining camps that traumatized his father with scenes of coffins, burning crosses, and searchlights on a mountainside in central Utah.

Bob and Shirley Thomas, circa 1949

Eight years after Zephaniah's death, Bob and Shirley moved their family back to Provo because Bob had been offered a position teaching English at Brigham Young University (BYU). Here at the school that had figured so strongly into his mother's happy memories, he worked his way into the administration, founded the honors program, and finished his career as an academic vice president.

I remember visiting my grandfather in his office in Kimball Tower, a thirteen-story office building in the middle of campus. My father had finished his Ph.D. in education and was teaching at BYU as well. In 1982, my grandfather retired to spend his time tending his garden and playing with his grandkids. My father moved up into the administrative ranks at BYU until he left to become a vice president at Utah Valley Community College. In 2001, when he accepted the position of president of the College of Eastern Utah (CEU), the Thomases returned to Carbon County after a seventy-five-year absence.

I am reminded of the image of the Castle Gate as I consider how my family history has circled back. Driving away from these cliffs, Zeph and Maud must have regretted leaving their friends. But there also must have been a tremendous amount of relief as those gates shut on their experiences of the strikes, the explosion, and the lynching. They believed it marked the end of their work in the mines, just as Evan and Margaret must have watched the seaport fade into the distance, thinking leaving Wales meant leaving coal. When my parents drove south from Provo past the Castle Gate to move their belongings and my younger siblings to the house on the hill overlooking CEU campus, mining reentered my family's consciousness.

During my father's eight-year term at CEU, my parents lived in Price, the largest remaining town in Carbon County. Most the former mining communities have been abandoned. Just as these towns appeared and swelled to capacity, once the coal was exhausted or no longer profitable to mine, the towns disappeared, following the four inevitable stages outlined by historian Homer Ashmann: prospecting and exploration, investment and development, stable operation, and decline.

Currently thirteen underground mines operate in Utah, ranking that state twelfth (at around 1.8 percent) in US coal production nationwide. Through modern technology, these mines yield four times as much coal as all of Utah's mines put together in 1925. However, most of the mines in Carbon County are now out of operation, and the remaining pit miners commute to neighboring Emery County. The majority of Carbon County's almost twenty thousand residents cluster in five cities; the county's population has grown by only a little over three thousand since 1925, when Zeph and Maud left.

In the far reaches of this county there are at least fourteen major ghost towns, as well as many smaller settlements and abandoned mine shafts. Historian and mine enthusiast Richard V. Francaviglia offers this explanation for the fascination of a ghost town: "The public understands a ghost town to be a tangible but depopulated place inhabited only by the memories of former occupants. The term 'ghost town' implies former activity, perhaps even former greatness, as manifested in decrepit buildings and other ruins. A ghost town is a place that is being reclaimed by nature, and we seem to take a perverse interest in the aesthetics and symbolism of time marching into, and over such forlorn places." On a drive through Spring Canyon, I spotted the remains of several abandoned mining camps. The foundations of stone buildings scattered throughout sandstone cliffs have become part of the scenery. Standardsville was named for its exemplary building design. Its five hundred residents lived in modern houses and steam heated apartments and enjoyed tennis courts, a baseball field, and a recreation hall. The cement retaining wall of the tipple still towers over the dirt road. Other than the scant remains of homes, the walls of the hospital represent the only other evidence of the community that sprawled out in the crevice between two mountains.

A half a mile up from this old mining community lies Latuda. A short jog from the road brought me to the skeleton of the mining office, once the largest structure in town. This building was intact until 1969 when a nineteen-year-old from Price planted half a case of stolen explosives in it and blew the building apart. He claimed he meant "to kill a ghost."

Erin Thomas, 2007

Mining office in Latuda

When the mine in Latuda closed in 1966, the last mining families moved lower into the surrounding communities, and around this time the legend of the White Lady became popular. Sightings of this ghost had been reported since the 1930s, when Latuda was an operational mining town. But an abandoned town added additional lure to teenagers from Price and the surrounding communities who hoped to catch a glimpse of her white, flowing robes. Reputedly a woman with a grievance with the mining company, she walks through the town almost to the threshold of the mining office and then disappears. There is no agreement on her history, except that she was a miner's widow. Some say an avalanche killed her child in 1927. Others assert her husband died in a mining accident, and the company never compensated her for his death. There is also considerable disagreement as to her final demise: hanging herself in the mining office, or dying in a mental asylum after killing her child to spare it from a life of suffering. My sisters, both of whom graduated from high school in Carbon County, drove up Spring Canyon with their friends late at night in search of this ghost.

Although not easily persuaded by hauntings, I felt tentative entering the ruins where a teenager in 1969 claimed to have found the body of a woman and chose to visit during the day. Broken glass scattered on the cement floor of the mining office next to the remains of a fire. It reminded me of the place under a bridge where teenagers in my town used to enact their mischief—still, after a quick inspection, I did not linger.

Locals still report seeing the White Lady of Latuda. One man, when questioned by a national ghost hunter, revealed: "I saw her once. Must've been back around 1985. I'd been prospecting, fiddling around really, in these canyons since I retired. I camped back here one night and while I'd hear all the stories, I never expected to see her. But around midnight, there she was. She walked past my fire like I wasn't there at all. I was too surprised to do anything but watch her go."

Not far down Route 6 from Spring Canyon is the turnoff to Scofield, the town below Winter Quarters. Here Route 96 cuts through the tops of mountains that, from a base altitude of over six thousand feet, look like hills. This mine is another abandoned shaft that is purportedly haunted; miners started reporting sightings a year after the Winter Quarters explosion. A 1901 article from the *Sun Advocate* reads: "The superstitious miners who are foreigners, have come to the conclusion that the property is haunted, inhabited by a ghost. Several of them have heard a strange and unusual noise, and those favored with a keener vision than their fellow workmen have actually seen a headless man walking about the mine and have accosted the ghost and addressed it or he ... Many supposedly intelligent men have claimed this and some twenty-five or forty have thrown up their jobs in consequence. These same people and others have seen mysterious lights in the graveyard on the side of this hill where many victims of the explosion of May are buried. Efforts to ferret out the cause have been fruitless ... These lights are always followed by a death ... Tombstones where the light appeared have been blanketed, but the light remains clear to the vision of those who watch from the town."

These days hikers and campers report hearing moaning and crying near the opening of the old mine, where the women waited in the rain for their men to be carried out in gunnysacks. When I turned down Route 96, it was also in search of ghosts, but there is more than one way to approach the dead. Scofield attracts not only those interested in the occult, but also those seeking communion with their ancestors. This sort of other worldly link is an accepted aspect of the LDS religion. Occasionally members will talk of the "veil being thin," meaning the veil between this life and the afterlife, this earthly existence and the heavenly existence. We believe in a realm of the dead where our ancestors have passed on only in body but maintain their individuality in spirit, and this realm is not far from where we are. In part, my belief in the living spirits of my predecessors motivates me to walk along their paths.

I have one early memory of Scofield. Nine or ten, I stood at an overlook, my eyes bouncing from each blare of red, orange, and blue sheet-metal roof that popped out of the dun colored summer landscape. I don't remember why I was there or who had brought me, only the connect-the-dot game I played with the colorful roofs in the valley below. More people lived there then than they do now; according to the most recent census, Scofield is home to twenty-eight people who live in twelve households in a town of seventy-eight households. Some of these are vacation homes that stay empty for the majority of the year. Others are just abandoned. In order to serve these twenty-eight residents, Scofield maintains five operating saunas, testimony to its lingering Finnish character.

These homes make up the few streets of a community wrapped around a railroad. A little white LDS chapel lies under the eastern mountains on a main street named Commerce. Across to the west, a graveyard is more populated than the current town. The last saloon marks the end of Commerce Street. Religion has lasted longer as a motivation for gathering than drink; although the church still draws a small crowd on the Sabbath, the saloon has been boarded up.

I arrived in Scofield on a fall afternoon when the hills were sprinkled with rows of gold-leafed quaking aspens, set off by the green-black of pines. I got out of my vehicle and shivered. The air had a cold, moist edge. I hesitantly jiggled the door of the saloon, unsure about entering if it yielded. I pressed my face up against the window to peer inside. Above the bar, a sign said "No Checks" and a clock stopped at 2:25. Stools stood vacant over a floor where the white tiles had begun to curl up like brittle sheets of paper. A cabinet was stocked with ancient bottles of Raid and Scope.

I stepped away from the windows and crossed back to my car to drive up to the graveyard. Scattered between the houses with colorful sheet-metal roofs, outhouses and boardedup shacks crouched that nobody had ever bothered to take down. A few feet from the end of the road stood a sign that read Open Range. Once a booming mining town, Scofield had reverted back to its original purpose as a place to feed livestock. Half a mile at the other end of Commerce Street, a tall, lean dog stared me down. It was the only living creature I had seen that day, but was strangely immobile—not turning its gaze or moving its tail.

There is no lush patch of grass in the Scofield graveyard because there is no gardener to tend it. Tombstones poke irregularly out of mountain grasses—wild, tall, and dry. Even so, in the disarray there are signs of remembrance. A plaque was erected in 1987 with the names of all the miners who died in the Winter Quarters explosion. The bronze relief above the names portrays a sense of torque, clamor, and night in the faces of coal miners who seem to be craning into a thick haze, the oil-powered lamps on their caps failing to light the path to escape.

At the dedication, Leslie Norris, former poet-in-residence at BYU and one of the most important Welsh poets of the post–World War II era, read a poem he had written for the occasion. I met Norris while I attended BYU. My writing teachers would invite him to class, to show him off, I suppose, but more because listening to him speak in his Welsh accent, with his careful gestures and vibrant imagery, was a beautiful experience. It was as if something dew-kissed and green-smelling had burst into the rooms and the white-painted bricks of the basement classrooms were breaking into verse. Once, in poetry class, he peered over the top of my poems and smiled at me. He was born in Merthyr Tydfil from mining stock, just like my ancestors. His father died in a cave-in there.

In 1983, he came to the United States to lecture at BYU for six months. He and his wife Catherine Morgan never left. I suppose he found the Mormon community under the rise of the Rockies a bit like the land of his countryman: the people cut off from the rest, a little peculiar, with a set of idiom all their own. His poem to the Scofield miners was published in a run of lithographs:

> I make this poem for the men and boys
> Whose lives were taken wherever coal is cut,
> Who went too early to the earth they worked in.
> I have brought with me to Winter Quarters
> Echoes of the voices of mourning women.
> … Let the men from Finland,
> The Welsh, the Scots, Englishmen, Frenchmen,
> Dying far from their countries a hundred years ago
> Let them be united in the rough brotherhood
> Of all tragic mines …
> I make this poem for the men who died
> When darkness exploded, and for their families
> And for those of us who come after them.

I walked to the back corner, where a marble obelisk engraved with THOMAS marked the resting place of Evan Jr. and Frederick. The stone stood six feet tall and was mounted on a two-foot piece of concrete. Red lichen grew on the base, and the inscription listed only their names and their dates of birth, followed by "sons of E. S. and M. D. Thomas." The first burial of a Thomas in the New World, it was executed under both the greatest poverty and the most pomp and ceremony. In the whole of Scofield cemetery, only one stone rose more imposing. Bearing the name of Edwin Street, the inscription had worn away, so that a couplet was barely legible: "To forget is vain endeavor. Love's remembrance lasts forever." An inscription similar to those on several other stones throughout the graveyard, this must have been a popular epitaph at the time. The few other massive grave markers in the cemetery belonged to Welsh and English families.

Most of the miners buried here were laid to rest in the front of the graveyard in rows. Each grave spanned wide enough for the coffin and a wall of earth between it and the next. Some families had replaced the original

Erin Thomas, 2007

New wooden headstones for Winter Quarters disaster 100th anniversary

wooden markers with six-inch cement markers engraved with initials, but many remained until 1999, when Ann Carter, a local resident, decided to beautify the graveyard to commemorate the one hundredth anniversary of the explosion. She and her husband Woody enlisted the help of the Utah State Historical Society and x-rayed the gray, splintered tombstones for the original lead-penciled names written by the blacksmith in 1900. Many of them had fallen over and split into shards. Now new wooden markers stand beside the old, engraved with names like Lasko, Kevlcaho, Kitola, Warla, Koloson, Heikkila, Jacobson, Maknus, Niemi, Bintella. In the front row of the graveyard, five small cement blocks mark the remains of the five Louma brothers.

The Finns in Winter Quarters kept their connections with the Old World. Later, when the mine closed in 1922, many moved to Salt Lake City, where a large Finnish community thrives to this day. They were among the two thousand people who came on May 1, 2000, to commemorate the death of their ancestors. The blast of a cannon at 10:25 a.m. signaled the beginning of the ceremonies. After graveside services, descendants shared their oral histories of the Winter Quarters explosion. Word had reached the great-grandchildren of Mataho Louma, the only son of Abe Louma spared in the disaster, who accompanied his parents back to their homeland. Two Loumas, a brother and a sister, came from Finland to participate in the celebration.

Ann Carter descended from a Finnish miner who came to Winter Quarters a month after the explosion to take the place of a dead miner. After I visited the graveyard, I sat with the Carters on their living room carpet, bending over a hand-drawn map of Winter Quarters. Scofield is located in a flat valley between the tops of mountain ranges. On the far southern side of town, a small road leads up to Winter Quarters. Tucked away and inconsequential, if it weren't for the accident, Winter Quarters would have disappeared from history. I would know little of the circumstances of Evan, Margaret, and Zeph in their first few years mining in Utah. Winter Quarters exists in the records only as a consequence of tragedy. Ann Carter would have never grown up in Scofield if it weren't for the accident that brought a new of wave of Finnish immigrants to the area. The Carters and I leaned over the map. Ann pointed to the schoolhouse, the superintendent's home, and the church.

My ancestors might have lived in Scofield or in Winter Quarters. In an area so small, it hardly seems worth making a distinction. In those days, however, it might have added a mile or so on to the morning and evening trudge to and from the shaft, where the miners would descend and walk another mile into the side of the mountain. Winter Quarters is private property now, and I unwittingly arrived in Scofield in the middle of hunting season. I drove to the edge of the town and parked in front of a gate. A stocky hunter in camouflage, with hard-cut Slavic features, was loading a four-wheeler into the back of a truck. According to an old farmer in a faded cap standing nearby, the hunter was the owner of Winter Quarters canyon. He ignored me for fifteen minutes while the farmer and his daughter warned me I might get shot. I insisted until the hunter turned and waved his hand. I walked up the road to Winter Quarters, convinced my pink and white striped shirt would set me off from the deer.

A farm snuggled into the mouth, but from there, the canyon thinned. A creek ran down the middle. Willows sunk their roots into the banks, and the woods climbed down the hill, covering where telephone poles used to carry electricity up to the mine. It was such a narrow canyon that I couldn't imagine how so many small wooden huts could have clung to the sides of the hill, the cold wind inserting itself through the cracks between the boards where the gum had worked free. I followed the path that Zeph

Courtesy of Western Mining and Railroad Museum

Ruins of Winter Quarter's Company Store, 1990

and his brothers would have walked every morning to the mine, trailing behind Evan. I know so little of this man, only that there was something in him that Margaret loved enough to spurn a rich man's son, and that he was short and tempestuous. There are a few foundations of outhouses on the outskirts of town, and some other lines of stones visible through the plant growth. The town was dismantled when the mine closed, every bit of stone and wood hauled out for other enterprises. Only two walls of the company store remain, cutting a lonely silhouette against the deeper canyon where the Finns and Greeks thrived, set away from the families of the other coal workers.

This intimate, wooded canyon seemed fitting as the ghost town of my ancestors. A deer ran across my path, and then another, in hunting season worse luck than black cats. I shivered, thinking of the hunters in the woods and the shot that could echo at any moment from the trees. I was so sure of my pink and white stripes, but watching the deer dash out of the underbrush diminished my bravado. The cool air held a light moisture, and a creek lined with willows trickled in the silence of the canyon. The colors of the forest deepened with the approaching dusk. I wasn't insensitive to the beauty or the danger; my heart beat loudly and my breath fell short as I climbed the trail of my ancestors. Fear, wonder, and love—the latter quickening my pulse the most. It was a grasping love of ghosts, spirits I wished would appear so I could say, "Yes, this is Evan and Margaret, and they lived here and I long to know them still."

In the photos of Winter Quarters taken just after the accident in 1900, it is early spring, and the quaking aspen are leafless. They make thin stick figures against the evergreens. At that time, the willows were cleared from the side of the river to make room for the railroad. In terms of trees, aspens have short life spans, somewhere between seventy and a hundred years, but the pines live longer. The pines that seemed to crowd closer into the slender valley as dusk deepened could be more than a century old. These same pines might have lined Evan and Zeph's early morning walks to the mine. These pines could have blocked the winter wind coming through the cracks in their cabin. It was possible that in this moment my life and the lives of Evan, Margaret, and Zephaniah were contemporary in the life of a tree.

Bridge

13

A Historical Gap

Since Zephaniah's death in 1943, no Thomas has spent a day's labor underground, but after moving back to Utah to work at BYU, Robert Thomas maintained his relationship with coal mining by taking weekend trips out to Sunnyside and Castle Gate with his family, driving past the five-hundred foot rock wall that divided his past from his present. This gate is impressed on my father's young memory and has become an emblem of my family's return to our roots.

Four Quartets by T. S. Eliot was one of my grandpa's favorite poems to teach his English classes at BYU, such that two of his former students quoted it at his funeral. In the last stanza of "Little Gidding" there is a reference to a gate: "We shall not cease from exploration / And the end of all our exploring / Will be to arrive where we started / And know the place for the first time / Through the unknown, unremembered gate / When the last of earth left to discover / Is that which was the beginning." My father believes that for my grandfather, this gate symbolized the Castle Gate, and my father referenced the poem in his acceptance speech at the College of Eastern Utah.

During these visits to the coal camps, my father witnessed the gritty details of a miner's life. He noticed the smallness and squareness of the homes, the piles of black coal, and the grime on the faces and working uniforms of the miners. In the stories my grandfather told his sons, he focused less on the accidents and day-to-day drudgery and more on the mischievousness of the mules and the idiosyncrasies of his coworkers. While visiting Castle Gate and Sunnyside, my dad was disturbed by the discrepancy between his own observations and the magic of his father's stories—only when he moved back to Price and came into regular contact with working miners did the magic come back.

The history between when Zephaniah and Robert left mining and the Sago accident that I saw on television in 2006 is difficult to bridge. Curiously, my family's shift from coal mining to education corresponds with a lapse in historical interest. Most coal mining histories end from 1910 to the 1930s; there is little contained in library volumes about the almost seventy years of its history since then. Only recently has an interest coal mining reemerged as it has come under attack by journalists who focus on its deleterious impacts on the land and the populace. There could be many reasons for this hiatus in record keeping.

After World War II, coal was no longer the fuel of choice; interestingly, the recent surge in writing about coal mining corresponds with a surge in coal production. Another factor could be the influence of family history on historians. Because mechanization has continued to reduce the need for manual labor, many have coal mining grandfathers, but few mine coal. Often authors refer to the coal mining experiences of their fathers and grandfathers who labored underground in the first half of the twentieth century and sparked their own interest in the industry. Additionally, some of the romantic appeal of the coal miner is lost with modernity and machines. Mechanization improved working conditions, and the nostalgic image of a miner catching a ride on a mantrip with pick in hand is no longer accurate. Today miners drive into the tunnels with trucks and manage equipment that digs the coal for them. Part of preserving the historical image of the coal miner may lie in disregarding his contemporary counterpart. Regardless of national attention, there has been much earth moved below ground during these seventy years, and after 1950, much earth moved above.

After the Depression slump, during which Zeph mined his last years, the World War II boom doubled coal production to fuel the factories supplying the military arsenal. Women joined the workforce, entering mines for the first time. In Carbon County, the population increased by a third between 1940 and 1950, and mining towns were hard pressed to provide housing for the newly arrived immigrant families. Only a decade later, the burgeoning market for solid fuel dropped off abruptly. The railroad engines that had been steam driven since their inception shifted to diesel, and natural gas and oil increasingly fueled the broilers of power plants. The company towns that burgeoned during the war, emptied. Maud had called them coal camps because she hoped mining would be a temporary occupation for her husband and sons, but this slight pejorative became prophetic as, one by one, the coal towns died off during the late fifties to sixties in Carbon County. In Price, the economy was in such decline that Governor J. Bracken Lee tried to close Carbon College, the predecessor of the College of Eastern Utah. Locals gathered fifty-six thousand signatures in sixty days, a number that exceeded the total population of

Emery and Carbon counties by at least twenty-three thousand, to keep it open. With the introduction of coal-fired power plants in Emery and Millard counties in 1978 and 1986, respectively, coal production in Utah increased over the midcentury decline. It now provides 83 percent of the energy used in electrical utilities.

Oil and natural gas beat out coal after the Second World War because they were abundant, cheap, and cleaner burning. Another advantage was the greater ease in transporting them over long distances. In 1956, uranium was first used peacefully to fuel the first nuclear power plant, and energy industrialists touted nuclear as the fuel of the future. But nuclear power enjoyed only a brief popularity until 1970 when concerns began to arise about the negative impacts of radiation on nearby communities and the difficulty of disposing nuclear waste—environmental concerns that were also voiced by the coal lobby. Economic viability posed another obstacle: building nuclear plants is prohibitively expensive. Finally, the partial meltdown of a reactor at Three Mile Island in 1978 halted further development of nuclear power plants. Oil also began to wane in popularity because prices were subject to political turmoil in the Middle East. During the 1973 oil crisis, the Organization of Petroleum Exporting Countries (OPEC) declared an oil embargo because the United States had supplied arms to Israel during the Yom Kippur War, and oil prices per barrel skyrocketed. The second oil crisis in 1979 resulted from the Iranian Revolution, when the Ayatollah Khomeini assumed power after the overthrow of Shah Mohammad Reza Pahlavi.

During the heyday of oil and nuclear power, coal leveled out at an annual growth rate of 1 percent, which had not changed since 1914, except for the World War II boom. During this time, the average energy needs of each American citizen increased dramatically, and coal fell from 56 percent of total energy consumption to 24 percent. Economists believed that coal was in an irreversible decline and would become obsolete as an energy source by 2050. The seventies signaled a turn of the tide. With the market failures of oil and nuclear, the yearly growth of coal doubled, rising to 2.2 percent between 1974 and 2005, a phenomenon dubbed the *coal rush.* Few coal plants have been built since the fifties, but the turn of the twenty-first century marked a return to investment in the coal-fired power plant.

Mining became more and more mechanized based on two inventions. The first was developed around the beginning of the twentieth century. The continuous miner is a massive tunneling and cutting machine that looks like a tractor with a giant drill bit on the front—the first was put to use digging the tunnel under the English Channel, a project started in 1870. In 1912, the Hoadley Knight Machine operated by cutting coal into nut-size pieces with an electric rotor. Improvements on these initial models emerged every few years, with greater horsepower and innovations to the cutting head.

Harold F. Silver, a Mormon from Salt Lake City, designed the first modern continuous miner in 1943. Learning of Silver's reputation for ingenious inventions in the dry cleaning and sugar beet industries, Carson W. Smith, the head of the Consolidated Coal and Coke Company of Denver, approached him about creating a device for safer, more efficient mining. Silver developed a machine, later improved on by the Joy Manufacturing Company, that removed coal from an entire face vertically using a cutting head, consisting of six revolving belts fitted with tungsten steel teeth, moving at five hundred feet per minute. As the head tore the coal from the face, caterpillar treads propelled the machine forward. A pan caught the coal, sending it along a conveyor belt to the bed of a shuttle car. *Newsweek* dubbed it a mechanical dragon; the *Chicago Tribune*, a giant red lobster; *Life*, a gargantuan mole. *Fortune* wrote:

> The machine in action is a wonder to behold; the ripping head sinks into the working face, rears and swings and bites again. The caterpillar treads crawl, the chain conveyor fills to overflowing, and a mountain of coal builds up behind the flexible tail.

In 1949, when the model appeared in the open market, it required a crew of four to five men to extract 250 tons of coal a day, rendering it 90 percent more productive than traditional drilling and blasting.

Continuous miners are used in both room and pillar mining and longwall mining, the two underground mining styles that persist to this day. In the room and pillar method, continuous miners are the primary excavating machines, but in longwall mining they are only used to cut the face before the longwall is installed. An invention first introduced to American miners after World Word II, the German developed "longwall" consisted of a plow of oscillating disks that ran along the surface of face, slivering a layer of coal onto a conveyor belt running below. This technique caught on slowly in American mines due to two major impediments: the cost of equipment (a longwall is a much larger piece of machinery than a continuous miner) and the laborious process of shoring up the roof. In the longwall method, miners initially used wooden props; around the beginning of the twentieth century, steel jacks kept the roof from caving in as the plow advanced. In 1961, at the Sunnyside mine in Utah, in the town where my grandfather Bob was born, Kaiser Steel Corporation developed a shearing machine that could remove harder coal from the face in a wider strip than the European plow. However, the technology wasn't used extensively in US mines until the 1980s, when self-advancing hydraulic roof supports were developed, reducing both the need for manpower and the safety risks associated with roof collapse.

As technology has progressed, the need for miners has continually decreased. For example, in Utah in 1996, 2,077 miners produced twenty-seven million tons of coal, a tonnage that was equaled in 2000 by just 1,672 miners. Mining technology companies are currently exploring Remote Ore Extraction Systems, where all underground operations are conducted by autonomous vehicles, drilling and cutting equipment, and sensors. Miners in control rooms would merely observe and manage the equipment. Already in working mines in Australia, load haul dump vehicles navigate themselves through underground tunnels with laser-scanning technology. Even if the United States continues to use coal, the underground coal miner may become obsolete.

During the period between 1940 and 2005, technology also improved mine safety, but only after three major accidents took the lives of over three hundred American miners. The Centralia disaster in 1947 is hauntingly similar to the Winter Quarters explosion of 1900. Built-up coal dust was ignited by a misfired shot, killing sixty-five in the blast and forty-six with "afterdamp." Miners had been complaining for years that the coal dust was not adequately controlled. Their claims were backed by a state inspector who had been reporting violations for five years, warning the repercussions could be deadly. An investigation after the fact revealed that the explosion had stopped when it reached the area where the mine was properly rock dusted, a process that uses pulverized limestone and other inert material to reduce the combustibility of coal dust.

Hearings convened in both the US House and the Senate to discuss mining safety. In 1910, the Bureau of Mines of the Department of the Interior was created to investigate mining equipment to prevent accidents and improve safety conditions. Inspectors did not have the authority to enter mines except when invited by the owners, and when they did make recommendations, they had no means to enforce them. John L. Lewis, president of the UMWA and known for his political clout and eloquence, lambasted the owners for their "criminal negligence," pointing to the statistic that mining deaths had superseded deaths in the armed forces during 1942. He continued: "If we must grind up human flesh and bone in the industrial machine we call modern America, then before God I assert that those who consume coal and you and I who benefit from that service because we live in comfort, we owe protection to those men first, and we owe security to their families if they die."

A man with heavy eyebrows and large jowls, Lewis ruled the union and bullied the government with a dictatorial hand. Like my grandfather, he was the son of Welsh immigrants and had worked in the coal mines as a boy, trying his hand at farming and construction as a young man until joining the labor movement in 1907. He led the UMWA from 1920 until 1960, becoming one of the most renowned labor figures in history, still beloved by miners across the United States. He also ranked as one of the most hated men of his time. Based on inches of paper, 5 percent of the *New York Times* in 1937 was devoted to John L. Lewis. His notoriousness grew during World War II, when Lewis disregarded presidential orders from Roosevelt and led half a million coal miners on strikes in 1941 and 1943, raising wages despite wartime freezes, so coal miners were the best-paid workingmen in America.

In response to Roosevelt's 1941 plea to "come to the aid of your country," Lewis retorted: "There is no question of patriotism or national security. If you would use the power of the state to restrain me as an agent of labor, then, sir, I submit, that you should use that same power to restrain my adversary in this issue, who is an agent of capital. My Adversary is a rich man named Morgan, who lives in New York."

In 1946, when the Truman administration had seized the mines in response to another strike, Lewis bargained with Secretary of the Interior Julius A. Krug to establish the Mine Workers Welfare and Retirement Fund, supported by a five cent royalty paid by the companies on every ton of coal mined; Lewis increased it to thirty cents by 1950. Lewis also pushed for a set of safety guidelines. The Federal Mine Safety Code for the Bureau of Mine Inspectors took shape in 1947, but like the 1910 law, compliance on the part of the operator was optional, and after a year of implementation this was estimated to be only around 33 percent.

It took yet another devastating mine accident to convince Congress that inspectors needed the power to enforce their citations. In West Frankfort, Illinois, 119 men died on Christmas Eve in 1951 after a pillar of gas ignited. Equipped with gas masks, fellow miners and the brothers and fathers of the deceased crawled through the debris in dust filled tunnels to recover the bodies of the dead. The corpses were laid out for identification in the local junior high school, awaiting their funerals, which were held on the 25th and 26th of December. A year later, with forceful urging from John L. Lewis, the Federal Coal Mine Safety Act passed under President Truman, allowing inspectors to close down mines they deemed unsafe. This law was still far from adequate; between 1952 and 1960, 3,179 miners were killed—approximately one each day.

In 1950, wearied by his frequent showdowns with the government and fearful of financial repercussions (Lewis was personally fined thirty thousand dollars between 1946 and 1948 by federal courts for ordering strikes), "labor's avenging angel" muted his strong rhetoric and adopted a gloved hand approach. With the demand for coal on the wane, Lewis could no longer threaten to shut down national manufacturing by ordering his union men to stay home from work.

By 1958, through closed door negotiations with major coal mining companies, Lewis had secured concessions of twenty-three to twenty-seven dollars a day in wages, forty cents a ton for the Welfare and Retirement Fund, and vacation benefits for members of the UMWA. In exchange, he worked relentlessly to put smaller mines (not incidentally, nonunion) out of business, assigning his vice president, Tony Boyle, to manage the strong-arm tactics perpetrated by union coal miners against nonunion ones.

As mechanization and market pressures continued to reduce the number of union members, Lewis developed financial power to replace his loss of manpower. He used monies collected in the multimillion dollar Welfare and Retirement Fund to purchase a share in the National Bank of Washington, and through transferring funds and buying out other banks, made this the second largest bank in D.C. In order to secure his bargaining power, he made secret loans to coal operators and railroads. To promote the use of coal, he bought into the electrical and bituminous mining industries. When forced to disclose its assets in 1959, the UMWA was assessed at $110 million. To his credit, Lewis doled out the Welfare and Retirement fund generously; in 1956, it provided seventy-five thousand miners' pensions, forty thousand widows and orphans' payments, and medical care for six hundred thousand patients.

When the depressed coal industry reduced payments into his fund, Lewis cut costs, reducing pensions and health care benefits. But when the market rebounded, Lewis used the excess funds to fight nonunion mines, rather than increasing payments to his retired and ailing members. For John L. Lewis, son of a Welsh coal miner, life had come full circle. The distance between him and his nemesis J. P. Morgan had narrowed, while the distance between Lewis and the men he represented had widened. In 1960, just as coal miners began to feel disenchanted with their champion, Lewis retired and nominated a more unscrupulous man to fill his shoes.

Tony Boyle is so demonized by the historical record that I've always imagined him stout and swarthy, with scars from youthful acne on a face that would redden with temper. Instead, Boyle was clean-cut and handsome—a very different look from Lewis's, who is often pictured with his thick white hair bristling above a scowl. Boyle also considered himself good looking; his first great "service" to the UMWA was commissioning large portraits of union leaders, the majority of himself. He hired his brother and his daughter, and set up his own pension fund. The UMWA's relationship with the coal companies continued under Boyle.

When the Consol No. 9 mine near Farmington, West Virginia, blew up in 1968, Boyle responded by conceding that "as long as we mine coal, there is always this inherent danger." He also claimed that Consol was "one of the best companies to work with as far as health and safety are concerned," although the mine had failed sixteen inspections. The cause of the accident that rocked West Virginians on November 20 was never determined. In a mine with 151 natural gas wells, the flames raged on for a week after the explosion. Twenty-one men had escaped, but hope of recovering the seventy-eight men still inside was abandoned, and the mine was sealed with cement to extinguish the fire. Almost a year later the mine was unsealed, so the bereaved could give their dead a proper burial. Recovery of the corpses spanned more than ten years. The bodies of nineteen men were never found.

The Farmington explosion was the first major mining disaster broadcast on national television, and people across America witnessed relatives of the trapped men express their hope of rescue and then deal with their grief throughout the ten days before the mine was cemented closed. The public outrage that arose from the incident had its effect, and a month after the accident, the Department of the Interior convened to discuss mining safety. The legislation that emerged in the aftermath, the Federal Coal Mine Health and Safety Act of 1969, mandated standards on safety as well as compensation for black lung. Callous to the needs of the men he represented, Tony Boyle teamed up with coal operators to oppose the law. The National Coal Association warned that its passage would force the closure of mines and cause a nationwide power shortage. In response, Representative Ken Hechler of West Virginia, one of the few politicians in all of history to back coal miners, did not mince words before Congress:

> The coal operators are shedding tears that this bill might force the closing of mines. They threaten power blackouts … We have heard this cry, the threat, this form of insensitive and implacable opposition for a long time. When are we going to place the priority where it belongs, on the cause of human life? When are we going to declare that the threat to close down a mine is not nearly as serious as the threat of closing down a man? The coal industry has fought every step of the way against measures to protect health and safety, and I am sorry to say that the union leadership has been insensitive to the needs of the men. There is no profit incentive in health, and there is plenty of profit incentive in high production, high wages, high coal dust levels, which produce a high rate of industrial murder.

High coal dust levels result in black lung, or coal miner's pneumoconiosis, a disease documented in the mining record since my Welsh ancestors first wielded a miner's pick. As early as 1837, miners in the United Kingdom had complained to the union that inhaling coal dust produced miner's asthma. Black lung finally became compensable in 1943, based on the findings of British physicians. American doctors rejected their research, claiming that their conclusions lacked evidence. This was despite the fact that 30 percent of all coal miners contracted the disease after working in the mines.

Three physicians In West Virginia, courageously opposed the opinion of the US medical community and traveled around the state, enlisting miners to support the proposed legislation. I. E. Buff, Donald Rasmussen, and Howley A. Wells would inflame the passions of their audiences by holding up a blackened lung in a Ziploc bag and saying, "Here is the lung of one of your comrades!" Wells would also take a dried piece of lung tissue and crush it to black powder in his hand. West Virginian miners formed the Black Lung Association (BLA), under local UMWA leaders Charles Brooks and Arnold Miller to advocate for their cause. Tony Boyle demanded that they disband; in response, the BLA introduced black lung legislation into the West Virginian Congress, and forty thousand miners went on strike to push it through. Despite Boyle and the coal lobby's opposition, on a national level the Federal Coal Mine Health and Safety Act passed, increasing aid to miners who had been suffering from black lung for years.

After six years of leadership, Boyle ran for reelection against Jeff Yoblanski, a dedicated union man who had supported the BLA. Exploiting his control over the union and the *United Mine Workers' Journal*, Boyle succeeded in throwing the election in his favor. When a federal judge called the legality of the results into question, Yoblanski and his family were murdered in their beds. The 1969 election results were ultimately thrown out, and in 1974 Boyle was convicted of ordering the murder of Yoblanski. Arnold Miller took over leadership of the UMWA, running on a reform ticket with the heavy charge of restoring the faith of the miners in the union while negotiating an upcoming challenge: coal industry meets environmental movement.

As the twentieth century progressed, the idea of the interconnectedness of man and nature gained ground, pricking the nation's environmental conscience. A poem printed in the *United Mine Workers' Journal* in 1950 demonstrates that the union was beginning to anticipate the effect that this would have on the industry: "We should practice conservation / We know we should not waste / the national resources of the nation / yet the greatest treasure wasted / is that of human life / more have death's sting tasted / while at work than in war and strife." The poem echoes John L. Lewis's indictment of the number deaths in the coal mines versus those in World War II, but this poem also pits conservation against the work of the miner, who had sacrificed much for the energy needs of the nation.

Surface mining began in 1950; to compete with oil and natural gas, the steam shovels and drilling equipment for coal increased in size and efficiency. By 1963, one-third of the coal mined in America was surface mined, producing twice as much coal per man-hour as coal mined underground. West Virginia was the first state to regulate surface mining in 1939 because locals had noticed that it caused soil erosion, increased the hazards of floods, polluted streams, ruined the possibility of agriculture, and destroyed other natural resources. Under this law, companies were required to obtain a permit for their strip mine operations and post a bond for reclamation of the land afterward. The actual repercussions of this legislation were questionable. Bessie Smith, an Appalachian woman from Kentucky, asserted: "Reclamation is a laugh. It's left up to the people to protect the land, help others retain their property and protect the beauty of the land." Accordingly, people in Appalachian towns began to assemble and fight strip mining through acts of nonviolent civil disobedience and sabotage of mining equipment. The Appalachian Group to Save the Land and People (AGLSP) was formed in 1965, and its members staged stunts, such as blocking bulldozers with their bodies or shooting at them with their hunting rifles. AGSLP was followed by Citizens to Defend the Environment (CODE) and several other grassroots societies. Although their primary concern in fighting against strip mining was protecting their homes and surroundings, some fought for their jobs. Among those in opposition were unemployed deep miners, a fact that caused Arnold Miller to take a nuanced position on the issue.

Strip mining employed miners in the UMWA, but it offered fewer jobs than deep mining. Reclamation, which was rarely enforced, also provided jobs, as the techniques that were more environmentally friendly tended to be more complicated and required more man-hours. Although the UMWA did not support the abolition of strip mining, it did support its regulation. By 1970, 50 percent of American coal came from strip mines, and 25 percent of America's mines were strip mines. Ken Hechler, who had fought to pass the Federal Coal Mine Health and Safety Act, also took a leading role in pushing legislation to abolish strip mining. In 1974, a law that would put strict environmental restrictions on strip mining passed in both the US House and Senate, but President Ford vetoed it on the grounds that it would cause hardship to the coal industry and raise energy prices. In order to show their opposition to the law, coal trucks from Wise, Virginia, made a forty mile caravan to D.C. with mottos such as "Strip Coal or Welfare Line?" hung from their sides. In 1978, Ken Hechler was voted out of office through efforts by the strip mining industry, and no Appalachian politicians have raised their voices to oppose the practice since, despite widespread opposition to the practice among the citizenry.

The passage of the Federal Coal Mine Health and Safety Act of 1969 was followed by the Federal Mine Safety and Health Act of 1977 and the formation of the Mine Health and Safety Association (MSHA) in 1978 to routinely inspect mines. Mining deaths have decreased, but injuries have not. In 1980, there were nineteen thousand recorded injuries, compared with twelve thousand in 1970. Mining safety's primary obstacle is enforcement: companies typically pay only 30 percent of the fines levied. Coal dust samples collected to determine the levels in the mine are tampered with by vacuums, leaving a legacy of one thousand miners per year still dying from black lung. Although the reduced number of miners working in mines combined with new safety laws have decreased the destructiveness of major mining accidents (in 1970, thirty-eight died at Hurricane Creek in Kentucky; in 1982, twenty seven died at the Wilberg mine in Utah), the numbers add up. In 1987, J. David McAteer, the Assistant Secretary of Mine Safety at that time, revealed that since the enactment of the Federal Coal Mine Health and Safety Act, 2,029 fatal accidents had occurred in coal mines. Adding the number of deaths through 2010, this totals 2,959. The all-time low of 18 deaths in 2009 was followed by forty-six in 2010. Perhaps the coal miner with black dust rubbed into the knob of his nose, navigating the dark tunnels of a mine with an open flame lamp, is not as far behind us as we would like to think.

West Virginia

14

The Little White Chapel

Living in the East, you get used to a consuming green—one that reminds you of what it must have been like for the first colonists who beat cities out of the tangled woods. I get the feeling when I drive around the D.C. beltway, where vines climb over the retaining walls, if we were to let the overgrowth go for just a few years, that nature would take back the roads. Roots would crack through the pavement, and it would seem as if there were never such a thing as a highway.

Seven months after the Sago mining accident, I drove through mountain roads in West Virginia, and this sense of green magnified. West Virginia is hill country. It is not the austere and jagged landscape of Utah, but untamed on the other side of excess: bursting, pollinating, chlorophyll-holic. The Appalachians are lower than the Rockies but are smothered in trees, and the ranges off in the distance fade into grades of atmospheric green and blue. West Virginia is what Wales must have been like prior to sheep grazing and mining in the days when the Druids built small shrines in the thick woods to the gods of nature.

Out West, a telephone pole just adds loneliness to a desert road. But here trees have been chopped lengthwise and widthwise to accommodate power lines that carve a path through the dense forest. The branches that remain remind me of the oddly sheared sheep in Merthyr. It is a hack job, a breach, and an interruption.

I followed US Route 220, a mountain road that wound between hills by the side of the Coal River, a tributary of the Kanawha River, along which John Peter Salley found coal in 1742. In West Virginia, settlers disregarded coal for many years due to the plentiful supply of wood. As in South Wales, the coal industry began small scale, gradually replacing charcoal in private homes. Around 1840, West Virginia received national attention, and the era of coal began.

The success of West Virginia's coalfields, like those of Wales and central Utah, went hand in hand with the reach of the railroad, and as rail lines extended, new fields opened. Slaves and local farmers mined it first, digging coal from the surface with picks and shovels. In the 1800s, following the Civil War, coal companies moved in from the surrounding states, built villages remote from town centers, and imported workers from the British Isles.

Although coal mining provides only 5 percent of the state's total employment, West Virginia currently produces 15 percent of all US coal—the largest amount next to Wyoming—with 277 working mines throughout the Appalachian Range. In Wyoming, mining involves the removal of only twenty to thirty feet of earth to access coal beds one hundred feet thick. West Virginian coal lies deeper and requires moving hundreds of feet of earth to get at veins only two or three feet thick. In order to stay viable in the market, West Virginia has cut its workforce and increased production. Despite federal regulations, economic pressure has caused unsafe practices to continue. West Virginia leads the nation in coal mining deaths, with 199 fatalities out of 617, or 33 percent, from 1993 to 2010.

A friend who offered me board at her parents' house for the weekend coached me on how to present myself in West Virginia. "Wear jeans and a T-shirt," she advised. "Otherwise they won't trust you." Weeks before my trip, I had arranged with Reverend Wease Day to visit Sago Baptist Church, located less than a quarter mile from the mine. At this little white church in the woods, journalists had captured the images of mining families waiting for news of their loved ones that had drawn my compassion at O'Hare Airport and propelled my investigation into coal.

Ten hours after I first saw reports of the midnight accident in West Virginia, a woman at Sago Baptist had received a call on her cell phone and yelled to the crowd, "They're alive!" According to Helen Winans, mother of trapped miner Marshall Winans: "Some damn person, I don't know whether they done it out of smartness or they didn't know what they was talking about came through the church a hollerin'." Family and friends immediately broke out into whoops and praised Jesus.

Down the road at Sago Mine, rescuers had penetrated the debris and were making their way through the tunnels. The fireboss, Richard Helms, had been found dead, killed from the blast. At nearly the same moment as the report of the rescue circulated among friends and family, rescuers reached the entrance where the crew of twelve had barricaded themselves against the carbon monoxide gas. Stripping away the plastic barrier nailed to the rock walls, the rescue team heard Randall McCloy's uneasy breathing and saw his chest move only slightly where he lay on the floor of the mine. One rescuer began immediate first aid care. The others gazed at where eleven miners leaned against the walls, much as if they were eating lunch, but did not move.

Among the twelve trapped miners, the crew had had only eight working *self-rescuers* (a small breathing apparatus containing one hour of oxygen) to share among them. Hopeful of being rescued, members of the crew had taken turns banging on pipes and roof bolts with a sledgehammer to signal their location. Some asphyxiated after eight hours of waiting, and some after ten. In their hands were notes they had written to their families as they felt themselves slipping. George Hamner Jr. addressed his daughter and wife: "We don't hear any attempts at drilling or rescue. This section is full of smoke and fumes, so we can't escape. Be strong, and I hope no one else has to show you this note. I'm in no pain, but don't know how long the air will last." The team carried out the lone survivor; an ambulance surrounded by state troopers drove him through the crowded streets of Sago and the nearby town of Buckhannon to the intensive care unit in Morgantown. The diagnosis was dehydration and a collapsed lung.

Meanwhile, the bells of the small white church rang into damp winter woods and people embraced. They had been keeping a vigil, praying two days for a miracle, and sang "Amazing Grace" to acknowledge God's mercy. Children danced in the aisles of the chapel. More false reports had circulated, and some were expecting the rest of the men to walk from the mine to the church, just ten minutes on foot, to see their families before they were taken to the hospital for examination. The crowd gathered inside started making room for the recovered miners in the pews. Around three in the morning, state troopers entered Sago Baptist Church and lined up along the walls. Mining chief executive officers (CEOs) and Governor Joe Manchin approached the rostrum. Friends and family members held their breath, expecting good news. Instead, mining representative Ken Hatfield announced that only the youngest of the miners, Randall McCloy, had been saved alive. Jubilation immediately exploded into rage; a fight broke out between the families, mining officials, and policemen. Ann Casto, whose cousin died in the explosion, expressed some of the feelings of the community: "[There are] some who don't even know if there is a Lord anymore. We had a miracle, and it was taken away from us." Nick Helms, the son of Terry Helms, who perished in the mine, expressed a different sentiment: "In reality they just lied. I immediately took my girlfriend, my sister, and the rest of my family out because honestly if they coulda got them, there wouldn't be anything of those guys left alive."

On the day I arrived in Sago, it was Reverend Wease Day's Sunday to preach in church. In our last phone call, he had indicated that his parishioners had recently received an upsetting report on the mining accident and wouldn't be disposed to interviews, but I wanted to see Sago for myself. Places like Merthyr Tydfil, Wales; Castle Gate, Utah; and Sago, West Virginia, draw little interest, outside of local historians until there is a landslide or an explosion. If it weren't for the Sago mining accident, I would have lived my life unaware of this small West Virginian town and the people in it. It was more than a coal-mining story that I was after, however. I wanted to extend and perhaps find compassion in a town that was experiencing events similar to those in my family's history. Mining explosions had shaped the character of the Thomas side of my family and may have contributed in a small way to who I was.

Reasons for the explosion in the Sago Mine are still being disputed. Just four days before my trip, West Virginia's government had released the initial reports. Lightning struck the ground at approximately the same moment that the methane in Sago Mine ignited during the early morning of January 2, 2006. The government inspector appointed by Governor Joe Manchin, J. David McAteer, designated this as the cause of the explosion. However, even he had his doubts—lightning struck more than two miles away from the mine, requiring it, according McAteer, to cross "the river, to the site of the mine … [and] go underground 250–300 feet."

Some of the locals saw the report as a "convenient" explanation, one that allowed the International Coal Group (ICG), owned by New Jersey investor Wilbur Ross, to dodge any responsibility for the accident. Sago Mine has had a history of safety violations: in 2005 alone it was cited 208 times for contravening the regulations of the Mine Safety and Health Administration, and some of the infractions could have contributed to the explosion. The report grated on an already sore nerve, as the locals had not forgotten the three hours in which ICG failed to report the news of the twelve miners' deaths.

Sago is nearest to the town of Buckhannon, where Reverend Day serves his other congregation. Though Buckhannon is the seat of Upshur County, it is a small community with only one main drag downtown, and on weekend nights only Maggie's, a sit-down restaurant, stays open past ten o'clock. I followed Day's directions carefully, driving on a road along the outskirts of town, passing a chapel, a John Deere machinery store, the local high school, and the junior high, until buildings thinned and I was heading back into the woods.

Off on the left, I spotted a narrow street called Sago Road. I took a quick turn, my tires screeching on the gravel. I followed the road built in the gully between two tree-covered hills next to a river and railroad tracks. Houses perched on both sides, reflecting different economies and time periods: some were of brick, others of grayed wood, some of aluminum siding. A large, tattered blue house with a long front porch reflected an old southern style—the whole wooden frame sagged and a "No Trespassing" sign was tacked on the front. The trees cleared

Erin Thomas, 2006

Sago Baptist church

suddenly; set back from the road at the end of a grassy lot was a beautiful white church, Sago Baptist.

Established in 1856, this church has had an influence on the community since its inception, serving the settlers that began to clear land and farm in the area in 1801. Reverend Day remembered the revivals he attended as a child—the flashlights that would burn from the woods as people would walk down from their homes to hear preaching about hell-fire and redemption. "Hell," according to Day, "had me so scared, I was afraid the Lord would get me before I got home." The gristmills were built around the same time as the chapel, followed by the sawmills. Farmers initially cut coal out of the hills to use in their stoves. Mining didn't begin in earnest until during the First World War.

The smallness shocked me. There couldn't have been more than twenty houses in the surrounding area. On the days Sago was in the national news, the hill on which the church perched had swarmed with reporters, microphones, and heavy camera equipment. On the morning I first visited Sago, a few cars parked in the gravel lot were the only sign of human presence. Dew lent a stickiness and sparkle to the quiet of insects, birds, and woods. The sun was just over the tops of the trees. Its rays refracted through the leaves, warming the top of my head. Gazing at the tree canopy piling up to a blue sky, I couldn't believe that anything bad had ever happened here. But it was July, and Sago Mine had blown in January. The sky must have been dark then, with the leftover broodings of a thunderstorm, the trees bare, and the air cool and hazy—weather for ghosts and tears.

Worship service hadn't started yet, so I walked up the hillside next to the church and read names on the old graves. Later, stepping into the lobby, I noticed, hung just above the chapel, an oil painting of a coal miner bending over a Bible, reading with the light of his headlamp.

A note scrawled on a scrap of paper was pinned to the bulletin board: "To the Sago Baptist Church and Pastor Wease Day. I was so touched by the generosity your church showed during this tragedy with the mine. Please accept this statue from a fellow Christian and a miner's daughter." It was signed by a woman from Clarksburg. Her gift was placed on the oak table below: a statue of Jesus laying his hand on a miner carved out of anthracite, the densest grade of coal—hard to ignite, slow to burn. The Romans were the first to carve it, digging it out of Wales and carrying it back to Rome, where they polished it into jewelry. They called it *jet*, missing entirely the full potential of coal and what it would mean to empire. While in Wales, I purchased a small statue of young mining boy in anthracite for my Grandma Thomas. It reminded me of my grandpa's stories of the mines in his youth, but especially of six-year-old Evan.

There was something intriguing about the image of a miner being carved from coal, and a Savior being cut from the same material to redeem him. This black rock has had so much influence in shaping our modern world, but also in shaping mining communities and the people in them. Coal is much more than a job; it defines an entire culture.

Next to the note on the board were plans printed from an email for a monument to be built to honor the fallen miners: Tom Paul Anderson, thirty-nine; Alva "Marty" Bennett, fifty-one; Jim Arden Bennett, sixty-one; Jerry Lee Groves, fifty-six; George "Junior" Hamner, fifty-four; Terry Michael Helms, fifty; Jesse Logan Jones, forty-four; David William Lewis, twenty-eight; Martin Toler Jr., fifty-one; Fred "Bear" Ware Jr., fifty-eight; Jackie Lynn Weaver, fifty-one; Marshall Cade Winans, fifty. Each of the miner's faces smiled, rendered in line drawings probably adapted from photographs. Some were wearing hardhats; others ball caps. The portraits were arranged in an oval to be etched by an artist into a stone slab and erected on the church grounds near the road.

A lean, medium-sized man with slouching shoulders and a gray mustache spotted me through the doors of the chapel as I browsed through the lobby. I looked away, suddenly self-conscious for snooping. He walked to the doors and opened one side to invite me in.

I thanked him and sat in the back pew. In the rows ahead, men sat in checked wool and flannel; their rolled-up sleeves rested on the bulge of their biceps. Their wives wore jumpers. Some had hair that was curled and permed, fluffed out in the style of the 1980s. Several I guessed were near my age, but they looked older, motherly. I hadn't followed my friend's advice about jeans. I was wearing a brown corduroy skirt and a T-shirt, an outfit I thought would be appropriately casual, but it was totally wrong for the congregation. My black, SoHo-style glasses immediately marked me as an East Coaster.

A flower wreath near the podium in the front honored one of the killed miners, sent to this small community by another concerned individual. Dick, the man who invited me in, started the service. The former lead elder, Martin Toler Jr., had been among the twelve killed in the mine. Dick first solicited the congregation for names to be included in the prayers. Members spoke out. Mumbles of assent and nods followed each request for remembrance:

"For Gus with kidney stones."

"Praise the Lord for our son with hepatitis, his enzymes come down."

"The Buhr family, their grandfather passed away."

"My neighbor's got cancer."

"And we need to remember our pastor, we should always remember him."

The Sago mining disaster hadn't been this town's first tragedy, and as I listened to each plea for help, I knew it wouldn't be their last.

Scraps of news reports and obituaries written by the bereaved came to mind: Peggy Cohen reported that her father, Fred Ware Jr., "was totally committed to the mines, has been since he was eighteen years old," and although he believed he would die in a mine, it never worried him. He prayed every morning that his coworkers would be safe. He left behind kids and grandkids, and his sweetheart, Loretta Ables, whom he planned to marry on Valentine's Day. Samantha Lewis remembered the way her departed husband, David Lewis, used to lace his boots at six o'clock in the morning, preparing to descend into the mine. He was only twenty-eight and planned to work in the mines as a roof bolter just until his wife finished her master's degree in health care administration. Jack Weaver's note to his wife told her: "I'll see you in heaven that's for sure. And Justin you'll always be my best buddy." His son Justin, when he heard the news of the accident, had grabbed his father's favorite cookie, clutching it in his fist as he and his mom drove down to Sago Baptist. Marty Bennett and George "Junior" Hamner came from families who had lived in Sago for generations.

Members of the congregation rose and knelt on the stoop before the podium, praying out loud as music played. The words "blessed" and "Oh Lord" rose and slipped back into the hum of voices.

Sunday school preceded worship service and focused on "the gifts of the spirit," a series of passages found in 1 Corinthians 12 in the Bible. Dick led the discussion. Members spoke out when they had something to add. "It's because of my faith that I am here," a man declared from the middle.

From the pews across came a reply, "But the Phillipses ain't here."

"They had a barbecue this weekend," one woman piped up.

"Yeah, they're probably really tired," another person mentioned.

"The beef was really good!" a man exclaimed, and several in the congregation chuckled.

Dick smiled and reeled the discussion back into Corinthians.

Occasionally, one of the older men from the back broke out into irrelevant pearls of wisdom: "God can do it, and if He does do it, it is to His glory. Godly people die every day, and somehow their dying is to His glory."

Toward the end of Sunday school, Reverend Day walked in, a large man with extremely white hair, and began to shake hands down the aisles. He had the look of a preacher: a broad smile; a large, sharply cut, triangular nose; and a potbelly that pressed out against his ironed white shirt where it was tucked into his trousers.

He shook my hand, asking, "This your first time?"

Children trailed in. There was only one young woman in the entire congregation, and she held the hand of the only young man. There were around seventy members in attendance, and I was struck by what an absence twelve men would create in a town like Sago. We sang an opening hymn. Ann Casto led—the cousin of Marty Bennett, who was killed in the mine. I had been singing every week in LDS congregations since I was a small child, but even this experience set me apart. I sang it straight; the congregation added twang. My music came out clean, and theirs feelingly.

Reverend Day took the stand. He announced that my grandfather was a coal miner and I was here to visit. Several people turned, and the men with rolled-up sleeves nodded their heads at me. Reverend Day led into his sermon in true revival style: "If I shut up, I'm gonna blow up." "First and most of all is love. It all boils down to love." Reverend Day worked in circles of reasoning; the logic wound and looped back in spurts. But as he talked, I felt the spirit—the warmness and pure light that accompanies sincere talk of deity.

Behind the reverend hung a tapestry of the Last Supper, and he picked up where Sunday school left off: "In 1 Corinthians 12 it talks about the body of Christ. Now the spirit is strong, but the body is weak. There is times when this ol' flesh is wore out. But when one member suffers, we all suffer with it. If there were fifty piano players and nobody knowed how to sing, what would you do? Plumbers, miners, all as good as the preacher. If you plugged your ears, you can't hear. You're all part of the family of Christ."

He stopped now and then to offer spontaneous prayer to "anyone who hasn't accepted Christ." I knew he meant me, the only stranger in the congregation. I had told him my Grandpa Thomas mined in Utah, and it was possible he'd concluded I was Mormon, or he might have just considered me a college-educated woman from the East, a relatively godless region. Either way, Reverend Day believed I wasn't Christian. He built up the emotion, prepping the congregation for a miracle. "I just preach as it comes to me," he admitted, and continued with stories of sick widows and funerals, careening back to the conclusion of love and being a witness of Christ.

Reverend Day grew up in Sago and moved back ten years ago when he was "called by the Lord to be a preacher." He drives a bus to supplement his income. He looks seventy, only twenty years younger than my grandfather would be. His salary was small—the church budget was read over the pulpit earlier in the service. As Reverend Day talked about his interactions with his parishioners, and the little white chapel he had helped raise money for and build, I understood why he had returned.

At the end of the sermon we sang "Amazing Grace," a song the reverend claimed to sing while he did his physical therapy. I noticed with some self-consciousness that the whole congregation could probably hear my voice, so different a melody it produced than their own. But this song was special to me, too. My family sang it the day we learned my older brother had terminal brain cancer, and although I sang it with my church voice, I sang it feelingly.

The reverend prayed again for the anonymous Jesus-less among them. Everyone knew he meant me. He asked the congregation to close their eyes. "Anybody who is ready to accept Christ, raise your hand," he commanded. I kept my hands folded loosely in my lap. He made one last attempt, "Erin, do you have anything to say?" Maybe these people needed some dramatic conversion: some good to come out of their loss, the reclamation of a stranger to the bosom of Christ. I wanted them to have their miracle, but I couldn't give him that.

"No, I don't." I answered.

A man with a handlebar mustache laughed out loud.

"Well, thanks for coming," Reverend Day conceded.

"Thanks for having me." I smiled back.

The service had reminded me of my own experience of loss—not in the sense of the sudden, but long, agonizing loss as my older brother died slowly from brain cancer. It had been over ten years since then, but I thought of him every time I set the table at my parents' home. We were no longer an even number, and leaving off the eighth plate always seemed wrong. When I visited Sago Baptist church, it had been a mere seven months since the accident, and for all twelve of the bereaved families, this missing plate was at the head.

As the congregation filed out of the chapel, I asked Dick where the mine was.

"Across the road, over the bridge, and just up a ways," he replied. "It isn't far."

I shook Reverend Day's hand as I left. As I climbed into my car, a man hollered after me, "Come again! We're just a bunch of hillbillies."

I drove down the road, over the railroad tracks and the bridge, and around a clump of trees; the mine was not even a quarter of a mile from the church. I could imagine how maddening it was on that day and night in January, the families waiting for news of their men who were so close, but unreachable.

The mine was merely a small bit of property surrounded by an electric fence. Heavy machinery lay at rest, although the mine had been under operation again for two months. There was a pile of coal at the end of a conveyor belt that descended down into a cavern in the side of the mountain. Although it was an eyesore in the middle of such lush greenery, it had little other perceptible effect on the surrounding area. Ironically, it is perhaps the history of deep mining that has much to do with the woodedness of West Virginia. Otherwise the trees would have been cleared long ago for timber or farmland.

I drove back down to the church and stopped next to an outline of stones that marked the foundation of an old house. I walked over to smell the pink blossoms of a single flowering bush. I was told an old woman used to live here; the house had burned down and nobody ever bothered to clear it. All that remained, aside from the foundation, was a tall, slightly tilting stone chimney. Nearby, a cement slab rested on the grass, waiting to receive the monument dedicated to the twelve Sago miners.

15

One Who Escaped

Typically the back windows of trucks in rural America bear items of erotica or patriotica, but on the back of Paul Avington's window was the decal of a miner crawling on his hands and knees with a headlamp lighting the way before him.

I followed the brown pickup to Paul's house. We had met up at a local gas station, either because he didn't trust an outsider or because he was afraid I'd get lost. "You Erin?" he had asked, sizing me up. My first impression of Paul Avington was that he looked benign, but I had never seen a man personify the word *burly* more fully. His frame was large and bearish, his graying blonde hair thick about his head, and his mustache fuzzy over his lip.

We walked into the front room of his home. His son shut off the television, and his wife asked if I wanted anything to drink.

"Pepsi?" she asked. "We've only got caffeine-free."

I wondered if the Avingtons were Mormon; few people shun caffeine otherwise. The possibility of sharing a religion with this enormous and whiskery man made me feel a bit more at home.

His wife gathered a few things, and his son scooped up a boy who was playing on the floor. "We'll leave you in peace," she said.

Paul was sitting in a large armchair near me and began to talk about himself in a very frank and easy way. He had been a miner for thirty years in seven different mines. Before Sago, he worked farther south, where the mines are unionized and the pay is better. I asked him why he became a miner, and his answer surprised me. His family had no history in the mines. "It seemed like a decent enough job. I went to school for a year in mining maintenance."

After reading the recollections of my ancestors, it intrigued me that anybody would choose mining as an occupation. However, as one woman in the Sago congregation had informed me at Reverend Day's service, with a mix of pride and amusement: "West Virginia is an interesting state. The four most popular occupations are health care, farming, mining, and driving coal trucks." Coal mining is among the more profitable. Although only 5 percent of miners have any higher education, an average miner's salary is around $55,000 annually, and miners can earn as much as $80,000 a year. Aside from the financial lure, there is apparently something addictive about mining. "If you work in the mines long enough, it's not blood but coal dust in your veins," I had been told by one miner, and a similar explanation runs through the accounts of many others.

Paul carefully explained current mining techniques. A cutting machine removes the coal from the face. A feeder scoops it onto a conveyor belt, where it is dumped into a shuttle car. The shuttle car drives it out from underground and dumps it on a stacker belt, which then feeds it up to a tower. At the precipice, the coal drops off onto a tall pile. Besides mining vehicles and buildings, the stacker belt, tower, and the mountain of pebbled coal are the only parts of this operation that are visible from the outside. As a shuttle car driver, Paul made fifty to sixty trips a day, carrying a load of eight tons on each trip.

In over two hundred years of coal mining history, there have been obvious improvements. Currently miners are trained in mechanics, hydraulics, machinery operations, and electrical systems. In order to set up a longwaller, miners use smaller tunneling machines until a seam of coal up to two miles wide is exposed. The longwaller runs the length of the wall on a track, scraping off coal by the inch to be transported from the mine. In order to maintain this process, miners run draining and ventilation systems and electrical alarm and communication systems. The only job that has changed little since the early days of mining is the work of the fireboss, who carries his gas meter into the mine and crawls through small spaces to check the levels of methane. Otherwise, mining is a far cry from the burden of the Welsh children who crawled through tunnels, dragging a tram behind them, and of little Evan, who loaded coal for his father. It is also distant from the way Zephaniah and Robert mined, blasting the coal with shots and breaking out the chunks with a pickax. Miners' lamps and hats no longer operate with oil and the open flames that so frequently ignited carbon gases. Even so, I asked Paul about his health.

Coal workers' pneumoconiosis, or black lung, is a broad term that covers a range of respiratory diseases acquired from the inhalation of coal dust, like the asthma Zephaniah Thomas died from in 1953. In 1969, Congress ordered an eradication of this disease. In 1977, the Black Lung Disability Trust Fund was created to provide compensations for its victims.

According to the United Mine Workers Association, 1,500 former coal workers still die from black lung each year. This number is not added to the death toll from mining accidents annually, which places coal mining as the one of the most hazardous occupations in America.

"I have an injured back," Paul confessed.

"From what?" I asked.

"Oh, bending over so long in tunnels. I hope it will last me ten more years. I'm fifty-five. I don't want to retire until I'm over sixty."

Paul's admission brought me back to Wales and Emylyn's stories of miners' superstitions about washing their backs—a strong back meant food on the table.

After some small talk, I broached the heart of the matter, "Can you tell me about the accident?"

Paul looked at his large hands and rubbed them together. He rolled his shoulders forward, humility taking over his large frame. I immediately recognized his sadness, his sense of helplessness and loss, but also something else. It was because of this question that he had agreed to speak with me.

The Sago Mine is shaped like a backward *F.* Two main corridors diverge perpendicularly from the main hallway: One Left (1L) and Two Left (2L). Each of these divides into a myriad of rooms where coal is extracted from the face. The thirteen men who died in the Sago mining disaster were working in 2L, the deepest branch of the mine, and most news reports mention only this crew. However, when the mine blew, there was another crew of fifteen men just ten minutes behind them heading toward 1L. This was Paul's crew.

Reporters had hounded the family members of the lost miners. Reverend Day had beat them off, while offering Christian hospitality in the form of cups of coffee and a seat in his congregation, just as he offered me. But nobody had come to Paul for his story. A miner goes into a mine every day knowing that there is some possibility of hazard and death. A mine groans, and this shifting and creaking of the earth may spark surges of adrenaline in a man who suspects a cave-in, but until a miner has experienced an explosion, there is no memory attached to this fear.

"My crew was just going down to our section on the main track. We threw the switch to go into 1L and then it exploded. Heat and dust and gas. It wasn't what you'd think; it wasn't a loud noise—the sound of an air poof." Paul made a gesture with his hand. He considered carefully as he spoke, trying to isolate the details of the experience from an overall sense of panic. "There was a roof fall. Then, in a split second, dust, debris, and heat. It didn't take us long, but we went through a lot of dust. I couldn't see my hand in front of my face, but I only thought of getting out."

He continued to look at his hands, and I could tell that not far from the relief, there was always the thought of the others. "I feel bad about the men who were killed. They tried to get out, but couldn't. The main fan was still on after it blew; we was closer to the intake air."

Owen Jones was leading Paul's crew in 1L that morning. When the air from the explosion hit the crew, it blew off his hat and shredded skin on another crew member's face. The carbon monoxide detector on his belt blared a warning that the men had only minutes to get out. But Owen's brother Jesse was working in 2L, deeper in the mine. Owen told the men that he was going back in, although Paul begged him not to. Owen joined the first rescue efforts and only turned back four hours later when it was apparent the carbon monoxide would kill him.

"How did you feel when you got out?" I asked Paul.

"I had lots of different thoughts," he said, crinkling his forehead. "I called my wife. It was the first time I been in an explosion."

Paul pulled out a collection of things from a folder and let them fall into his lap. There was an issue of the *United Mine Workers Magazine* with an article about the deaths. He sorted through pieces of paper and articles dealing with the accident, pulling out ones he thought would interest me. He fingered a medium-sized patch and held it with reverence: "In memory of our fallen Sago miners, January 2006," it read. When I asked him about the reasons for the explosion, he related the story of the lightning. There was no bitterness in his voice; he did not suspect ICG of covering up the causes.

16

A Memorial

The woods were wet and splotched with a pale green fungus. Rust and yellow colored leaves clung limply to the ends of branches; there had been no snow, and remnants of fall still littered the forest. A mist hung low over Buckhannon River, where one stone-block pillar stood like a solitary omen. This pillar once supported an old mining bridge for railroad tracks. A local had told me that she used to play in the river during the summertime, but in early January, no children laughed or splashed. The water lay perfectly still.

It was just over a year after the accident, and I stood on the bridge overlooking Buckhannon River across from Sago Baptist Church. The area looked as it must have then—the edge of sun over the hills touched the moisture in the air with cool light. The tree cover that had looked so thick and protective on my first visit in the summer was completely bare. Straight trunks lined up one after another, appearing smaller as they climbed the hills on either side.

A week previously, friends and family members of the twelve Sago miners had walked across the bridge to the mine to pin black ribbons and black and white roses on the fence. It was the third gathering at the small white chapel, and public attention—as in the wake of all tragedies—had begun to diminish. The event was unofficial and private; two dozen people participated.

The public memorial service had been held in August in front of a much larger crowd. Men and boys had set up folding chairs on the grass under canopies from the local funeral home. A stage was erected on the rock foundation of the widow's house, and the tilting stone chimney finally dismantled. On the day of the ceremony, the sky rumbled, threatening to

Stone block pillar in Buckhannon River

pour rain on the heads of the mourners gathered on the grounds of Sago Baptist Church to view the memorial: a six foot slab of granite, reflective and dark like the walls of a coal mine. Many came wearing T-shirts with the names of their deceased loved ones printed across them, and this time fewer camera crews camped along the sidelines to record the events.

Governor Joe Manchin spoke to an audience of several hundred. He had been there with the families when they waited for news of their loved ones. Like many West Virginians, Manchin has a connection to coal and even to its tragedies, as his uncle died in the Farmington explosion of 1968, which left seventy-eight dead. Family history counts him in, but in the Sago disaster, the worst since Farmington, some feel he has failed to support the miners' families. Newly established safety regulations have been slow to be implemented in a state that lost twenty-six coal miners in 2006 in onsite accidents. Although investigations and conferences have been held to determine the cause of the Sago explosion, locals feel the results are far from satisfactory. Some among the miners must have grimaced at Manchin's words. President Wilbur Ross of ICG and other company representatives sat in folding chairs placed in front, looking out over the audience of mourners. I wonder if they shifted in their seats just a little.

When Reverend Wease Day and Reverend Roger Foster (another preacher in the Buckhannon area) took the stand, the muscled arms folded across the chests of the miners must have tightened with emotion. The women might have sniffled and sponged the tears off the faces of their children with wadded tissues rummaged from purses. These families would recognize those who had truly done battle for them. Concerning that night months before, at the first gathering at Sago Baptist, when they had waited in anticipation for the thirteen miners to walk across the bridge to meet them, Reverend Foster said: "Folks, they didn't cross the Buckhannon River, but they did cross a river, and as Christians, we call it the River Jordan."

After the names of the fallen were read, the names of those who survived were listed: Paul Avington's crew, as well as Randall McCloy, who was able to attend and be honored with the rest. Rescued from the mine in a coma, he had made steady progress and, after weeks of treatment, was able to eat for himself and open his eyes. By the time of the service, he was walking again, although he had some disability on his right side and permanent visual impairment from the hours with little oxygen. Later, he attended the small gathering to mark the one-year anniversary of the explosion.

Because this mining community recovered only one of thirteen, they have clung to him. West Virginians across the state have followed his recuperation closely, and his wife, Anna, received the Mother of the Year award from a nonprofit organization dedicated to the preservation of state history. For many, he is an emblem of continuing grace—that despite tragedy, there are still evidences of mercy. The street he lives on was officially changed to Miracle Road. Later, Sago Road was renamed Coal Miner's Memorial Roadway, a ponderous title for a sudden left into the woods.

Since his recovery, Randall has tried to give back to the mining families who lost their fathers and husbands. In April, four months after the accident, he wrote a letter addressed to "the families and loved ones of my coworkers, victims of the Sago Mine disaster," where he detailed the last moments of his comrades:

> We eventually gave out and quit our attempts at signaling, sitting down behind the curtain on the mine floor, or on buckets or cans that some of us found. The air behind the curtain grew worse, so I tried to lie as low as possible and take shallow breaths. While methane does not have an odor like propane and is considered undetectable, I could tell that it was gassy. We all stayed together behind the curtain from that point on, except for one attempt by Junior Toler and Tom Anderson to find a way out. The heavy smoke and fumes caused them quickly to return. There was just so much gas.
>
> We were worried and afraid, but we began to accept our fate. Junior Toler led us all in the Sinner's Prayer. We prayed a little longer, then someone

> suggested that we each write letters to our loved ones. I wrote a letter to Anna and my children. When I finished writing, I put the letter in Jackie Weaver's lunch box, where I hoped it would be found.
>
> As time went on I become very dizzy and light-headed. Some drifted off into what appeared to be a deep sleep, and one person sitting near me collapsed and fell off his bucket, not moving. It was clear that there was nothing I could do to help him. The last person I remember speaking to was Jackie Weaver, who reassured me that if it was our time to go, then God's will would be fulfilled. As my trapped coworkers lost consciousness one by one, the room grew still and I continued to sit and wait, unable to do much else. I have no idea how much time went by before I also passed out from the gas and smoke, awaiting rescue.

I crossed back over the bridge to the monument below the white church. The faces of the miners had been etched into the smooth stone, according to the plans I had seen on the church bulletin board months before. It was capped with a miner's hardhat, and the bottom was carved with wheels, suggesting an old-style coal tram. Above the wheels was the quote: "We'll see you on the other side." It was adapted from the note that the former first elder of Sago Baptist, Martin Toler Jr., wrote to his family. Toler was a mining foreman with thirty-four years of experience and, according to McCloy's account, prepared to lead his crew into the next life with the Sinner's Prayer, an Evangelical acknowledgement of God's grace and the willingness of a sinner to be saved. The men must have been like those I met at Reverend Day's service, only with their faces covered with coal dust. Perhaps some of them took off their hardhats and placed them over their hearts for respect:

> Father, I know that I have broken Your laws and my sins have separated me from You. I am truly sorry, and now I want to turn away from my past sinful life toward You … I believe that Your son, Jesus Christ, died for my sins, was resurrected from the dead, is alive, and hears my prayer. I invite Jesus to become the Lord of my life, to rule and reign in my heart from this day forward. Please send Your Holy Spirit to help me obey You, and to do Your will for the rest of my life.

Two benches flank each side of the monument, and one is placed directly across from it, with the images of Randall McCloy and the Sago Baptist Church engraved on the top. The other two are engraved with the names of the sixteen miners of the second crew, Paul's crew, who exited the mine alive before the camera crews and rescuers arrived. This was Reverend Day's idea, who realized these men had been overlooked in the ensuing publicity after the explosion. Some felt honored to have their names on the monument; others felt that it was a precursor to a tombstone.

Below the stone tablets were artificial flowers with small drops of rain from the night before on their petals, as well as other mementos left for

Erin Thomas, 2007

Base of sago memorial

the dead: an evergreen wreath tied with a red bow addressed “Dad we love and miss you so!” signed by Peggy and Daniel, and a small name plaque for “Jerry” next to a bouquet of red roses. A hat with “Mine Rescue 2007” written on the brim with fabric marker snuggled close to the granite stone, nearly covered by the pots of flowers. Most curious was a porcelain angel in a blue clingy dress, with roses springing from her arms and gown.

In our society, erecting monuments is a means of reconciliation: to both honor and commit ourselves to remember. But living without a father or a husband is a daily experience. After the ceremony dedicating the monument in August, the widow of George Hamner Jr. said: “I think the monument is just beautiful. But what would be better is if my husband were alive today. We shouldn’t be here today.”

In addition to the twenty-six deaths in West Virginia in 2006, there were twenty-three more nationwide. The coal industry hit a ten year high in mining fatalities, which was ironically matched with record highs in coal production and profits. Just the day before I traveled out to West Virginia to visit the monument, 2007 had its first mining accident with the loss of two more miners due to a roof collapse in Cucumber, a mining town on the southern tip of West Virginia.

Lightning has remained the official story for the Sago Mine disaster. This, according to many West Virginians, is a way of dismissing responsibility to "an act of God." Many harbor skepticism about how the lightning was able to enter the mine after striking the ground two miles away, and they favor the alternative explanation. The day of the explosion was the first day back after the holidays. The fans had been switched off for four days, which is illegal, according to federal regulations. Notably, on the day of the accident when there was still hope of rescue, government inspector McAteer was more concerned about the possibility that flipping a switch could have ignited the methane gas than the lightning strikes that morning.

Near the time the memorial was dedicated, three families filed suits against the ICG, asserting that negligence had caused the accident. Although Sago Mine is not unionized, the United Mine Workers of America has responded to these accusations with a report of their own, stating that sparks generated in the mine by falling rocks or by rocks striking against roof supports were a more likely cause of the accident. The report also blamed the miners' deaths on the lack of effective communication systems installed in the mine, a direct result of ICG's money-saving policy.

More troubling than the lack of resolution and the persistent questions in this case are the two miners that took the blame upon themselves. John Nelson Boni was the fireboss of Sago Mine, responsible for checking the levels of gas. Five days before the accident, he alerted his supervisor that he had detected a small amount of methane that had leaked into the mine. Boni's report stopped there. His supervisor was not concerned. William Lee Chisolm was a dispatcher in charge of alerting crews of high levels of carbon monoxide. An alarm sounded twenty minutes before the mine blew and Chisolm, following procedure, alerted the crews within the mine to verify it. After the accident, Boni retired and Chisolm took a leave of absence. Boni had worked in the mines for thirty-six years, and Chisolm for eleven years. Although neither was accused of any wrongdoing pertaining to the accident, the constant questioning by investigators gnawed at their consciences. Chisolm started drinking heavily, then shot himself in late August. A month later, Boni followed suit with a bullet to his head—two casualties of the intense media attention focused on the disaster.

A month after visiting Sago the second time, I heard that Sago Mine had been closed down in response to the report of the UMWA. I thought about Paul Avington on the day of our interview, that large, gentle man filling an entire easy chair, saying he'd like to be able to work ten more years. Since our interview, he reported that his daily shuttle car trips of eight tons of coal each had become tedious. Coal mining had become tainted by the memory of his fallen comrades and his own sense of neglect. Still, Paul needed that job. I thought of Evan and Zephaniah Thomas, who were never successful at any other job they tried. Now in our specialized society, expertise is even more crucial in the workforce. Who would hire a fifty-nine-year-old coal miner?

Sago Mine, with its numerous safety violations over the past few years, should be closed. And Sago isn't the only one. In the United States, we currently count yearly mining mortalities by the tens, but in developing countries they bury miners by the hundreds each year. In 2006, just over a month after the miners were trapped in Sago, sixty-six miners awaited death, trapped a mile down in a Mexican mine. Months later in China, which has recently become notorious for unsafe coal mining practices in its push for industrialization, fifty-seven miners were drowned by underground flooding. Early in 2007, not long after two miners were killed by a roof fall in Cucumber, West Virginia, one hundred workers were killed by an explosion in Siberia. Throughout history, mining has been characterized by the sacrifice of the smallest and weakest.

As a coal miner's granddaughter whose ancestors suffered in the industry, I'm inclined to spare future generations from the darkness of a mine. When I think in terms of the process of moving to alternative forms of energy, I imagine the industry dying out as miners like Paul retire, all receiving their benefits and a pat on their burden-bearing backs. Yet, I know better. Change is a process of closings, a process of firings that leaves individuals jobless and without resources. There are always families.

Reading the reports, I found that Sago Mine had only been idled, and the miners who worked there had been given jobs in other mines operated by ICG. I took a deep breath of relief for Paul. But I couldn't help think of the town of Sago, a place where the main industry is mining. After the explosion, ICG had already begun to reduce its workforce, cutting the number of underground workers from ninety to forty-four by the end of 2006.

According to Pam Campbell, the sister-in-law of Marty Bennett, who died in Sago Mine: "We basically fight for what we feel needs to be done for the rest of these coal miners in this area. Because as you know, coal mining is a very, very big part of this town. We need coal mining, but we need safe coal mines."

In line with this sentiment was a discussion string on the Internet site "Volition Forums." Janjojo, Missy1970, Maylinda, and LisaBGoesShopping closely followed the events unfolding from the Sago mining disaster. All four women are West Virginians and related in some way to mining. Through their correspondence, the whole saga of the explosion unfolded: the reports in local and national media, the money-making schemes masquerading as donation drives, scholarships at Wesleyan University for the children of the deceased, and the recovery of Randall McCloy. An article *Vanity Fair* published that was critical of the coal industry in West Virginia generated quite a bit of traffic between the four women and other sympathizers. Wrote Missy1970: "About it saying the impact on the land. Not just because my husband is a coal miner, and that is how we pay the bills, but I look at it this way. This is God's Land, and He gave someone the knowledge to know how to mine coal or it wouldn't be done. Further more God knows what he is doing to the land by giving people the knowledge and know how of mining coal. Sounds like a bunch of tree huggers wrote that article."

The views this coal miner's wife attributes to tree hugging would probably not be far from my own, and I think of a conversation I had with my father while driving through Price Canyon. In central Utah in 1999, the German mining company RAG AG opened the Willow Creek mine right next to the Caste Gate shafts—the tunnels crisscrossed underground. Knowing the history of the mines in the area, and that they were apt to be gaseous, the company took every safety precaution to ensure ample ventilation. There was a cheerfulness that set into Price and Carbon County, seeing coal being piled once more on the hillside. For them, it meant miners at work and economic prosperity. Only two years later, methane escaped into the mining tunnels and caught fire. A well-trained rescue crew was able to evacuate all the workers except for two.

After the explosion, mining efforts were abandoned and the shaft closed up. The mining equipment has never been recovered, and to this day a longwaller, a piece of machinery about five hundred feet long, is stalled in the position where it was abandoned in six years ago. In line with the strict reclamation laws in Utah, the site has been reseeded with sagebrush, the buildings removed, and the opening of the mine filled in. From the road it is impossible to see any difference between the mining grounds and the surrounding foothills.

As my father explained this to me, I felt a sense of great satisfaction as we passed by the one remaining, monumental wall to Castle Gate. Other than the graveyard, there are no visible remains of the mining camp my grandfather grew up in or the subsequent mining efforts, only pale brown shale, brush, and the magnificent rise of the mountains. It was as if the negative impacts of the explosions, the strikers and strikebreakers, and the burning crosses of my grandfather's childhood fears had been erased.

However, after the Willow Creek mine closed, Price's economy went into a nosedive. It took years for it to recover. Government officials and energy specialists in the area are still analyzing the remaining coal reserves in the belly of that reclaimed piece of mountain and considering the possibility of cutting it right back open.

To disappear, to be assimilated back into the woods; to have Paul's shuttle car rusted and overgrown with vines, the continuous miners shut up in the hillside under the room and pillar maze of 1L and 2L, Miner's Memorial Road cracked by tree roots, the blue house drooping into a

Greg Smith

Sago Mine, 2007

pile of old boards sprinkled with pale yellow fungus. This is not what the people of Sago want, to have their style of living eradicated.

The last time I visited Sago, I left Reverend Day on one of the memorial benches, his face shaded under a straw hat. The foundation of the widow's house and the bush had been uprooted, and next to the monument a new home was being erected, still just sheetwall and board. Fred "Bear" Ware's son had bought the property to be close to the monument. From the church, I drove to Sago Mine and walked up to the gates surrounding the complex. Inside, a woman sitting under a pavilion bolted to her feet. "You can't take a picture here," she ordered. She looked afraid, and I suddenly felt sorry for her.

Sago Mine had been opened nine years previously by Anchor Coal. After a short period of operation, it was closed when a helicopter dropping an employee off at the tipple crashed into the side of the mountain, killing the president inside. Other companies had deemed the mine unfit to work, but ICG reopened it. The surrounding community was grateful, because it meant jobs. ICG later closed the mine, after putting operations on stall for months—people talked too much of ghosts. Farther down the seam, ICG opened another entrance to the tunnels inside, renaming the mine Imperial in hopes of erasing the negative associations with the accident. This mine cuts into the mountainside under the Alban cemetery, where both of Wease Day's parents are buried. Near the cemetery, lives his father-in-law, a retired miner who went to work every morning thinking that he might not come back one day. Now, ICG is tunneling right under his home.

17

Mountains Made Low

The New River is one of the oldest rivers in the world and one of the few, other than the Nile, that moves from south to north on the globe. It curves through mountains covered with thick greenery that slants up on either side, and in the spring and summer months professional river guides steer groups of four to twelve tourists straddling bright colored inner tubes through the rapids. After I left Reverend Day near the mining memorial in Sago, I drove south to meet up with some friends in Fayetteville to take a trip down the river. Before my group set off, my friend Greg asked a group of guides why they had chosen their occupation. I was expecting them to mention the danger and the rush of adrenaline or the beauty of working outdoors. One nicknamed Chicken shrugged his shoulders: "You can be a river guide or a coal miner, there's not much else."

I dug my paddle into the current. Feeling the push against the muscles in my arm unleashed something gutsy and raw. *I am in mountain country.* My body awakened. *I am alive*—this thought reasserting itself with each stroke. I remembered how, along the River Napo in Ecuador, the guide had let me ride through the rapids bronco style, gripping a rope on the front of the raft and balancing with my arm raised high over my head. He had turned over the guide's oar and let me steer our group through a section of the river. One of the older women crouched low in the tube because she was afraid I would tip us. Chicken had a wild look in his eye, but I knew not even he would turn over the paddle to me—too much liability.

The landscape along the New River was surprisingly similar to the River Napo, tawny colored boulders and trees wound with vines. Forty feet up from the river, a railroad track was just visible, carrying cars full of coal from sites beyond the mountains. The smoke rising from the engine

The New River

seemed lyrical on first glance, but the mines from which they traveled were likely strip mines.

Company owned roads block pleasure-seekers who come to camp or raft in West Virginia from witnessing anything but a pristine image of the land. In such dense forest, it takes only few miles of wildness to cover the destruction from the mountaintop removal that is in process behind. This, according to activist and local resident Julia Bonds, is West Virginia's "best kept dirty secret."

The New River is like the River Napo in more than just appearance. Ecuador has begun to market jungle tours and rafting to protect its rainforests against the intrusion of oil companies who drill, dig, and send pipeline through the forest, a process that makes it impossible for the local natives to continue their way of life. This sort of industrial invasion is something I tend to associate with communities in developing countries, not in the United States. But in both West Virginia and Ecuador, it is either eco-touring or fossil fuels—there isn't much else in terms of larger-scale economics. The primary jungle around the River Napo is reputedly the most biologically diverse rainforest in the world, and the growth that blankets much of the mountains and hills in West Virginia is the most diverse hardwood forest. Classified as mixed mesophytic, these woods support over eighty species of trees, where most others sport only two to three. The foliage that blankets West Virginia is the oldest in the United States, and some ecologists believe it is the "mother forest" from which tree species of the eastern temperate zone have spread.

Mountaintop removal (MTR), a practice opposed by two-thirds of West Virginians, begins with clear cutting the tree cover, decimating the plant and animal species that live within. The bedrock is blasted with explosives. In West Virginia, this averages three hundred pounds a day, creating a much larger detonation than the shots used to blast coal in my grandfather's day. Clouds of coal dust rain down on towns; the foundations of houses and wells crack; boulders are launched into the nearby communities, and occasionally through roofs. After the earth is broken up, draglines—tractors up to twenty stories tall with buckets capable of scooping one hundred tons of soil—remove the *overburden*, the technical term for up to six hundred feet of mountain that lie between the miners and the coal seam, typically not more than a few feet thick. The removed soil is dumped into valley fills, blocking streams and poisoning water downstream in a state where a third of the residents still draw groundwater. Employing fewer miners and operating above ground, MTR is much safer than underground mines such as Sago, extracting coal with fewer immediate human casualties. Historically, graveyards mark mining sites where all other indications of mining activity have ceased, reminding us of the human cost of carbon energy. In the case of mountaintop removal, the graveyard is the lasting impacts on the land.

Appalachia is a Native American word for "endless mountain range," a term used to describe the land as well as the folk culture that inhabits it. Over two hundred years old, this culture was first developed by either the Scotch-Irish or London's outcasts—the tale varies—and was added to by later immigrations of Germans, Poles, Italians, and other ethnicities recruited by the mines. Being *removed*, or separate from the mainstream, characterizes Appalachians, who are independent and close to their neighbors and the land that surrounds them. Although in some ways estranged from the rest of America by their culture, location, and mistrust of outsiders, they self-identify as Americans more readily than any other group of people in the United States.

Beaten brass stars adorn many houses snugged into mountain valleys. Emerging off a mountain road, I spotted a tin roof painted to resemble Old Glory, a proclamation of patriotism to any fowl or aircraft above. At a campground a week before the Fourth of July, my Appalachian neighbors had already begun their celebration with music, laughter, and fireworks into the wee hours of the morning—my humble plea for taking it down a notch resulted in a drive-by bottle hurling directed at my tent as the party broke up for the night and an injunction to "go get a cappuccino." It was possible I had violated some regional cultural norm of courtesy. In any case, I was an outsider telling them how things should be done. In the context of the many years of abuse by coal companies, I understood why they were wary of strangers. In this case, the fatal flaw of the Appalachian people has not been mistrust, but trusting outsiders too well.

Marfork Hollow, Blair, White Oak, and Kayford Mountain are names of towns destroyed or partially destroyed by mountaintop removal. In these towns, fathers passed their land on to their sons, and their sons built homesteads, generations living side by side, preserving their traditions of living off the forest. Planting, gathering, and hunting follow a yearly schedule. In winter lands lie fallow. March is planting season and brings wild greens, maple syrup, and ramps. Ramps, or wild leeks, are harvested in the spring with such local zeal that they have their own festivals in towns across the Appalachian Range, where plates are piled high with ramps and other local foods, such as potatoes, fried apples, beans, and cornbread. Summer is berry and bass fishing season. Falls brings the harvest not only of gardens, but also of wild nuts—hazel, hickory, and butternut. "Gathering" in the Appalachian forest is usually for personal use, but there are a number of products harvested for profit; among them is wild ginseng, called *seng* by locals.

Ginseng is a plant that has never been domesticated and must grow wild or in circumstances that simulate a hardwood forest. It takes five to seven years to mature and will only grow under specific conditions of shade and moisture. Those who can follow the signs of Solomon's seal, cohosh, wild ginger, snakeroot, and spleenwort to a seng plant poking up out of the mulch, sporting its red berries, can earn an average of $450 per pound of dried root. In 1994, the year for which the most recent statistics are available, West Virginia exported a total of 18,698 dried pounds. American ginseng is highly sought after in China due to its cool *yin* quality, and collectively the industry earns seventy million dollars per year in exports.

But, like many indigenous cultures before them, the Appalachian people are being driven off the land. Although state policies of supplying natives with diseased blankets and outright extermination are considered shadows of the past, the coal companies are backed by nearly every politician in West Virginia. In towns as close as one hundred feet from MTR sites, periodic dynamite blasts crack the cement foundations of houses and wells. Coats of dust accumulate on porches, and once clear mountain streams swirl darkly with coal sludge, topsoil, and silt from the overburden. Behind homes, the forest that once supported the seasonal gathering of wild honey and berries is obliterated. After the surrounding natural areas are destroyed, the coal companies often go after the towns. Some residents finally sell out, and some of the more stubborn are driven out by the courts, which in large part support the coal companies. In 2006, Patriot Coal wrangled with courts to buy out the Caudill family property hiers, who lived in Laurel Branch just outside of Hobet 21, which at the time covered over twelve thousand acres and was the largest strip mining project in the world. When some members of the Caudill family refused to sell, Massey contended that "coal mining was the highest and best use of the property."

The superlative *highest* reveals language usage about coal particular to West Virginians. Appalachian folk have a deep sense of spirituality. Among West Virginians, 86 percent identify themselves as Christian; consequently, in terms of what should or should not be done with mining, God's stance is invoked frequently. "The good Lord put coal in these mountains for us to take advantage of it," claims William R. Raney, president of the West Virginia Coal Association, as if this is the final word. But this is not what many of the residents believe. Judy Bonds compares the physical rape of the land to a spiritual rape: "These mountains are in our soul. That's what they're stealing from us. They're stealing our soul."

Mountaintop removal began just a few years after 1971, when the West Virginian legislature, spurred by broad citizen support and activism, nearly outlawed strip mining. This "strip mining on steroids" was developed as the most cost and labor efficient method at getting to the deep deposits of low sulfur coal in the Appalachians. Since then, fifty thousand acres of land have been decimated, 1,200 miles of stream covered, and 2,500 peaks destroyed. Environmentalists predict that in twenty years, half of the mountain peaks in West Virginia will be leveled.

Political protections against MTR consist of the Clean Water Act of 1977 and the Surface Mining Reclamation and Control Act of 1977, both passed under the Carter administration. The Clean Water Act prevents industries from dumping waste into waterways and requires them to obtain a permit from the Army Corps of Engineers for any kind of "fill." The Mining Reclamation and Control Act, commonly referred to as SMCRA, requires companies to obtain permits before surface mining and to place a bond to cover the cost of reclamation. According to this law, the land is supposed to be restored to an "approximate original contour"—defined as within fifty feet of the initial elevation—and to be stabilized with grasses and shrubs. Streams are also protected by the stream buffer rule, where no land can be strip mined within one hundred feet of a stream.

Throughout the 1980s and 90s, companies largely ignored these regulations, and inspectors turned their heads until a lawsuit was filed by the West Virginia Highlands Conservancy against the Army Corps of

Engineers for giving permits to MTR companies who dumped waste into streams. A federal judge backed the environmental group's complaint, thus threatening the existence of all MTR sites in the Appalachians. At this point, the Clinton administration stepped in: the environmentalists agreed to drop the lawsuit in exchange for the promise of tighter federal scrutiny over mining projects.

On February 2, 2005, President George W. Bush proclaimed the following before Congress and the nation: "To keep our economy growing, we also need reliable supplies of affordable, environmentally responsible energy. Nearly four years ago, I submitted a comprehensive energy strategy that encourages conservation, alternative sources, a modernized electricity grid, and more production here at home—including safe, clean nuclear energy. My Clear Skies legislation will cut power plant pollution and improve the health of our citizens. And my budget provides strong funding for leading-edge technology—from hydrogen-fueled cars, to clean coal, to renewable sources such as ethanol. Four years of debate is enough: I urge Congress to pass legislation that makes America more secure and less dependent on foreign energy."

Listening on the radio, I latched onto the expressions "environmentally responsible energy" and "alternative" and was surprised to find myself in a rare moment of agreement with the former president. Examining the statement later, I saw that little pointed to what I considered alternative. "Alternative" and "environmentally responsible" meant one thing: coal.

During the 2004 presidential race, the coal lobby donated $1.5 million to the GOP, and the Bush administration faithfully chipped away at the protections in the Clean Water Act and SMCRA by adding exemptions and altering words. The refuse prohibited in the Clean Water Act was clarified as "fill," paving the way for companies to gain permits to dump the overburden into streams. The stream buffer rule was emended to suggest that companies use "the best technology currently available to prevent, to the extent possible," polluting streams when they fill them. Companies are currently allowed to dig ditches to replace the covered streams, washing silt and toxic chemicals downstream, where fish have been discovered to contain high levels of poisonous selenium.

The diverted runoff has caused flooding that, during the particularly bad flood years of 2001–2, killed fifteen West Virginians and damaged over $150 million in property. The reclamation required by SMCRA usually consists of spraying hydroseed: a mixture of seed, water, fertilizer, tackifier (glue), and green wood fiber mulch. This is just a step above green paint—concerned locals assert that hydroseed would sprout on a picnic table. A popular treatment for commercial and roadside areas, hydroseed precludes the possibility of trees ever growing. Although it has been suggested that the topsoil, dumped bottom first into valley fills, be saved to be spread back over reclaimed land, companies argue that this solution is too expensive.

Under SMCRA, due to the impossibility of truly restoring the MTR sites to their original contours, mining companies were offered a blanket exemption, provided they converted the land to public use. Over the past thirty years, out of MTR's fifty thousand leveled acres, reclamation efforts have culminated in the regional contribution of flat land for two high schools, two golfing greens, a jail, an FBI complex, and a hardwood flooring plant. Blair Gardner, legal counsel for Arch Coal, describes these reclamation efforts in typical West Virginian rhetoric: "I believe in God, so I am reluctant to say we are improving what the Creator made. But more pertinent is, are we doing something inherently good? I think we are."

According to the Bush administration, the loosening of restrictions on MTR was consistent with a policy that sought to gain American independence from foreign oil, ensuring a steady supply of cheap, domestic energy. And although the Obama administration has promised to regulate mountaintop removal, it has not yet removed the wording changes in the Clean Water Act. Consequently, powering the nation's energy needs come at the sacrifice of one of the America's most unique endemic cultures and the self-sufficiency of some of our poorest citizens. But, as one of West Virginia's top coal CEOs reportedly explained, lounging in his office chair behind a sheet of bulletproof glass, it all comes down to the bottom line.

West Virginian politicians know this well. During his terms of state governor, Joe Manchin changed all the welcome signs on the borders of West Virginia from "West Virginia. Wild and Wonderful" to "Open for Business." Coal profits from mountaintop removal only benefit a privileged few—1 percent of West Virginia's workforce makes a decent living as a result, but the bulk of the cash lands in the hands of the legislature and out of state companies.

In 1981, the Appalachian Alliance released the Appalachian Land Ownership Study, which revealed the startling level of displacement in land ownership: "Of the 13 million acres included in the survey, nearly 75 percent of the surface acres and 80 percent of the minerals were absentee owned. Forty percent of the land and 70 percent of the mineral rights were held by corporations—mostly coal." Because state governments in the Appalachian region discount property taxes for extractive companies, 53 percent of this land, which is held by the companies, generates only 13 percent of the property taxes. This results in underfunded schools and municipal services, as well as communities with high poverty rates, low median incomes, low high school graduation rates, and high levels of substance abuse.

"In simple terms it is either fatalism or the coal industry," Appalachian sociologist Helen Lewis asserted about the cause of poverty in this region. The poverty of extraction-based economies has recently caught the attention of economists, and Harvard University researchers Jeffrey Sachs and Andrew Warner identified a relationship between natural resource-based exports and the decline of GDP in 1995. The only way that this trend can be mitigated is by an investment in education and a transparent political process, but dirty politics practiced on a population with relatively low educational levels is leverage that West Virginian coal barons have been unwilling to give up.

Massey Energy Company CEO Don Blankenship is a West Virginian boy, raised and educated in the state. Like the Crawshays of Merthyr Tydfil, he rose from humble origins and built himself a "castle" on top of a local mountain from the spoils of his industry. Combined with its subsidiaries, Massey Energy is one of the largest coal companies in the world and dominates the coal business in West Virginia. Although he is the highest paid coal executive in the country, Blankenship, unlike the Welsh ironmasters, has built no schools or hospitals. Instead, Massey's culpabilities include having the worst fatality rate for miners and the top ranking for violating environmental standards.

In order to preserve his position of power, Blankenship vacationed with West Virginia's Supreme Court Justice Elliott "Spike" Maynard and poured $3.5 million into a fake nonprofit, "And for the Sake of Kids," to unseat Justice Warren McGraw, who ruled against Massey Energy. Journalist Michal Tomasky wrote that Blankenship was "famous in West Virginia as the man who successfully bought himself a state supreme court justice in 2004 and then tried to buy himself the state legislature, failing spectacularly at the latter effort." Few civilian lawsuits have won against the coal baron.

The Army Corps of Engineers, the Surface Mining Board, the Department of Environmental Protection, and the Environmental Protection Agency (EPA) have been lax in defending the law, while the federal government has chipped away at their environmental protections. Appalachia remains one of the least educated and poorest regions in the United States, and MTR robs the people of access to the knowledge of the mountains, which has sustained them for more than two hundred years.

To the rest of us who neither mine nor profit from coal, coal means energy. The industry often claims that the United States is the "Saudi Arabia of coal," touting the statistic that we have sufficient resources to last for 250 years at current energy consumption rates. When we speak of the energy crises, in terms of coal we'll most likely make it though the next century. Excluding the concerns of climate change and mining mortality rates, there is still one major obstacle to exploiting this national bounty. This coal, threaded in seams through the Appalachians and the Rocky Mountains, lies under natural areas and towns. It lies under people's lives.

18

String-Town Appalachia

String-town Appalachia was a barrage of small white signposts that popped up on the left side of the road before I was conscious of crossing a town line. I was on my way to Sylvester, the first stop in my journey to the sites that had made headlines in the fight against mountaintop removal. Seth, Kirbyton, Fosterville, Orgas, and Keith charged by in ten miles on West Virginia Route 3, Coal River Road. The smaller of these towns indicated they were unincorporated, meaning there was either too little interest in forming a township or too few citizens to lead it. The residents paid their taxes to the state and identified themselves according to the county they lived in. Few side streets cut into these towns, which were hardly more than an interruption between the road and the sheer rise of forest stacked on forest above. This growth nudged its way back into civilization, as evidenced by telephone poles covered with vines that graced the road like leaf-wound crosses.

Earlier that morning, traveling from Beckley to Racine, I had counted one store, seven gas stations, twenty-four churches, four coal operations, and ten flea markets. The latter outcrops of industry were laid out on cars and lawns: old shoes and clothes, empty Coke and beer bottles, ceramic figurines, knives, and rusted ax heads. In a strip of towns with so few mercantile institutions, some of these products had likely made many stops up and down the highway, bought and sold in rummage sales throughout the years. Each time a flea market peeked out of the trees, I craned my neck from the driver's side to snatch a glance, but the river-cut path hurled me around its contours. No time for pausing, insisted the locals, tailing me although I was doing 50 mph.

Erin Thomas, 2008

Silver dome at Elk Run preparation plant

Coal River Valley, an area so seemingly untouched by modernism and development, is the battleground for much of the recent debate over coal mining and its most destructive practices: water and air pollution, sludge lagoons, mountaintop removal, and flooding caused by diverted streams. Appalachian townspeople miles removed from grocery stores have recently found themselves in the forefront of the conflict, broadcasting their stories to a national audience. Sylvester is one such town, and I knew I'd arrived when I caught the almost metallic glint of an enormous white dome behind the trees. I pulled off into the neighborhood just under its shadow. A woman walking her two dogs asked me if I was looking for something.

The dome covers the pile of coal adjacent to the Elk Run preparation plant, a facility installed in 1998 by Massey without regard to direct requests from the town mayor, the town council, and a petition signed by 75 percent of the residents to locate it elsewhere. There is a second step between the mine and the power plant. Before coal is burned, it must be cleaned to remove the toxic chemicals. Many times these toxins are leached into the surrounding groundwater. Aside from the water pollution, there is also the sound of the coal's delivery, the hum of the facility, and the danger to locals from coal trucks steered by red-eyed operators at corner cutting speeds. The townspeople of Sylvester were most concerned about the winds that traveled east in the area, sure to deposit residue on their homes, providing both a health and an aesthetic hazard. Within a month of plant's operation, the fears of residents proved true as coal dust accumulated on their windows, their porches, and the exterior of their houses. Their complaints went unheeded for two years by the coal regulatory organizations until Bill Cook, a member of the Department of Environmental Protection, decided to stick his neck out and issue violations to the plant, a gesture in West Virginia that gets somebody reassigned fast. In response, the Elk Run facility made superficial efforts to reduce its impacts on the town. The most effectual, a sprinkling system, was only operated when inspectors were in town.

Pauline Canterberry, Mary Miller, and other town members realized it was going to take more than just complaints to force Massey to cave. They filmed the coal dust that covered their cars, clogged the screens on their windows, and billowed in black clouds over their homes. They then presented this to the Surface Mining Board, which unanimously voted that the plant was violating its permit and ordered the company to correct the problem. Massey appealed. Residents realized that more would have to be done, so they filed a lawsuit against the company for damages to property. For two and half years, Pauline and Mary gathered dust samples, earning themselves the title of the Dustbuster Sisters. They reached out to the media and local environmental organizations to bolster their cause. In the end, Massey erected the white dome, the triumph of an Appalachian town.

The woman with the dogs told me she was the environmental protection engineer at Massey, and I asked her opinion of the company's dealings with the town of Sylvester.

"I don't agree with everything that Massey does, but eventually they'll put it back the way it was. Eventually, they'll get there."

"I used to think the last place I'd work was the coal industry, but there's nothin' like the pay with Massey," she admitted.

She claimed that the dust levels were an exaggeration, and that Massey had installed dust meters on every street. As an example, she pointed to her home located right under the shadow of the dome. "You can see how close we are. My house would be black."

The local pastor of Sylvester Baptist Church, David K. Minturn, begged to differ. "Massey says we don't have a coal dust problem. It's better than it was, but we do have a coal dust problem. I swept my porch this mornin' and it's black. Road dust in't black. We have a town meetin' every month, but not much is done."

The town's fight for justice caused rifts in his congregation when Massey threatened the job of a church member who worked for the company. Minturn understood the importance of the industry to the area. Both his daughter and son-in-law work for coal companies. "If it weren't for coal, the young wouldn't stay and buy houses here. These towns would disappear when the retired people died." But there is more than one way to run a company, and Massey, in Minturn's opinion, violates the second law of Christianity: "The Bible says there are two laws. Love God and love your neighbor. If an employer obeys these laws, he treats his employees well. You got a company like Massey and the union won' have anything to do with it, so it's like a law unto itself. If you don't fight them, they don't change."

After their victory against coal dust, the town pushed for more reforms—lowering speed limits in order to prevent roadway deaths caused by overloaded coal trucks, removing an illegal dam in the Big Coal River that runs through the valleys alongside Route 3, and preventing an additional plant from being built at the Elk Run facility. The townspeople of Sylvester realized that reform ultimately meant getting involved on a larger scale, so they took their story to news conferences in Washington, D.C. and on road tours through Minnesota, Ohio, and Pennsylvania. They joined in support of neighboring communities in their battles against King Coal and raised money for local environmental organizations. In West Virginia townspeople must petition regulatory organizations to follow the laws already on the books. Sylvester went one step further, campaigning for new protective laws and for politicians who were not bought by the coal companies.

West Virginia's coal barons accuse "crazies" and "greeniacs" of providing the momentum behind the grassroots movement for environmental justice, while it would be more instructive to examine the backlash of their own practices. An example of this is Chuck Nelson, a retired miner who worked for thirty years in the mines, whom I met at an anti-mountaintop removal gathering. He had worked for Massey Coal for five years after they bought out the previous owner. Not long after the company moved into town, Massey decided to build the Elk Run preparation plant. Chuck began to mobilize the residents of the city, even though it would probably mean losing his job. He later quit to avoid tarnishing his work record. For his activism, Chuck was blacklisted by Massey and has been unable to find work with any other company. Frustrated with his resistance efforts, company executives finally approached him: "What is it going to take to get rid of you?" Chuck is a graying man with a mesh faced ball cap on—mid-height, mid-weight, in no way an imposing figure. "You could have settled this a long time ago if you'd been willing to sit down with me," he told them. Now he works alongside the local environmental organizations.

One of his key concerns is Massey's waste impoundment that looms four hundred yards behind the Marsh Fork Elementary School near the town of Sundial. The earthen dam is 385 feet tall and holds back 2.3 billion gallons of murky liquid. Dust is not the only concern surrounding a coal cleaning plant, as the chemicals used in the process are known carcinogens. Heavy metals such as mercury, chromium, cadmium, boron, selenium, and nickel are some of the impurities washed out of the coal and dammed up in what has come to be known as a *slurry pond*, or *sludge lagoon.*

Chuck's concern is not unwarranted. In 1972, a slurry pond built by the Pittston Coal Company collapsed after two days of heavy rain. In minutes, 130 million gallons of sludge poured into West Virginia's Buffalo Creek, sweeping away homes of Logan County residents, with the residents still inside. The final toll amounted to 125 dead, one thousand injured, and four thousand homeless. No indictments were made. The disaster, according the company, was "an act of God."

In 2000, an impoundment owned by Massey burst near the city of Inez in Kentucky. 250 tons of slurry spilled into nearby streams, creating a twenty-mile dead zone and polluting the water systems of ten counties. Massey, like Pittston, blamed it on the rain.

More recently, in December 2008 the containment wall at the Tennessee Valley Authority power plant in Kingston breached, spilling five hundred million gallons of waste over four hundred acres of land. Fifteen homes were damaged by the spill, but the residents escaped unharmed.

Although there are few outside of the coal industry who would assert that the incident near Inez was an act of God, it could be interpreted as a warning. Miraculously, there was no loss of life in this spill, but children at Marsh Fork Elementary in Sundial may not be so lucky. The landslide in Aberfan, Wales, is an obvious parallel: children and teachers pressing their faces against windows, a double line of white arches, *a silence in which you couldn't hear a bird or a child.* When experts estimated that those at the Marsh Fork Elementary would only have three minutes to evacuate, Massey responded by installing a blow horn.

The sludge lagoon in Sundial is the wastewater depository for the Goals Coal Processing Plant, a Massey subsidiary, and dust from the operations, proven to cover every surface in the elementary school, is being inhaled into the students' lungs. In 2009, 250 children attended Marsh Fork, and many missed school for problems with asthma and stomachaches that were a likely result of the waste impoundment nearby. Petitions to government officials went unheeded, so local citizens banded together in the Pennies for Promise campaign to raise the funds to build a new elementary school. In 2006 Ed Wiley, whose granddaughter attended Marsh Fork Elementary, walked 455 miles from Charleston, West Virginia, to Washington, D.C., to raise awareness of the cause and convince local politicians to act on behalf of the children. After walking forty-three days, he reached the capital, where he met with West Virginian Senator Robert Byrd, who later sent him a personal letter acknowledging that Marsh Fork Elementary presented "a dangerous situation." Ed Wiley and his cohorts continued their Pennies for Promise campaign, earning funds slowly while the children continued their studies under the shadow of a sludge lagoon.

The town of Blair is another hotspot in the fight against MTR and the historical site of coal mining's most significant debacle. Located in

Logan County, which neighbors Boone County where Sylvester lies, Blair is twenty miles into mountain country past a number of unincorporated string towns. Along Route 17, the flow of residences is steady until one approaches the Blair city limits, where homes begin to thin, separated by large patches of lawn spread between. Past two white chapels and across from an abandoned store, I found what I was looking for. Alongside the road, posted with no more ceremony than the town signs before it, was a cast bronze plaque marked The Battle of Blair Mountain, commemorating the culminating event in the ten-year-long Mine Wars:

> In August 1921, 7000 striking miners led by Bill Blizzard met at Marmet for a march on Logan to organize the southern coalfields for the United Mine Workers of America (UMWA). Reaching Blair Mountain on August 31, they were repelled by deputies and mine guards under Sheriff Don Chafin, waiting in fortified position. The five-day battle ended with the arrival of the US Army and the Air Corps. UMWA organizing efforts in southern West Virginia were halted until 1933.

Mining wages had plummeted after the end of the First World War, and miners struggled to unionize the coalfields of West Virginia to ensure their wages. These attempts were countered by the Baldwin-Felts Detective Agency's private agents, hired thugs who operated under the sanction and funds of the mining companies, that in turn operated under the sanction of many leading officials of the state's government. A showdown was sparked between Baldwin-Felts agents and local law enforcement in the city of Matewan in southern West Virginia, when the agents unlawfully turned unionized miners out of their homes. The ensuing gunfire felled seven detectives, two miners, and Cabell Testerman, the mayor of Matewan.

During the fight, Sid Hatfield, the chief of police, shot Albert Felts, the brother of Thomas Felts, who founded the agency. Months later, while Hatfield and his wife were approaching the McDowell County Courthouse, Hatfield was gunned down by Baldwin-Felts agents before he reached the steps. His assassination was the impetus for the Blair Mountain uprising. After months of living in the squalor of tent colonies, the evicted miners were only too eager to rally against an establishment that offered them no fair recourse.

With red bandanas tied around their necks (by some accounts, this coined the term *redneck*), miners marched toward Logan County, singing: "Every little river must go down to the sea / All the slaving miners and our union will be free / Going to march to Blair Mountain / Going to whip the company / And I don't want you to weep after me." Sheriff Chafin and his men had cleared forest and dug trenches at the peak of Blair Mountain; miners approached through the underbrush, dodging bullets from the machine guns and rifles above, but making little progress due to the thickness of the cover and the constant stream of ammunition. A day later, the federal government gave in to Chafin's plea for help and sent infantry troops and twenty-two planes from the Army's Air Service. When the federal army began to arrive by train, Bill Blizzard called off the assault.

The miners believed the intervention by the national government would bring needed attention to the conflict and an answer to their concerns. Caching their guns against future conflict, the striking miners returned home, waving their flags and expressing confidence in Uncle Sam. Immediately after the Blair Mountain battle, however, membership in the UMWA dwindled by the thousands across the United States. It was not until twelve years later that the National Industrial Recovery Act of 1933 granted all miners the privilege of joining unions without retribution. In terms of casualties, the impact of this battle in the Mine Wars is difficult to determine. Estimates of the number killed at Blair Mountain range from twenty to fifty; no official count was taken because the miners packed out their dead.

The roadside sign is the only visual memorial of the Battle of Blair Mountain, an event that marks American history as the largest citizen uprising since the Civil War, and the only time the US government sent the air force out against its own people. Beyond, Route 17 climbs to the summit of Blair Mountain, which rises gently upward. Here among the ferns, pods, and rot that accumulates thickly, providing fertile ground for wild ginseng, there are bullets and signs of conflict still preserved from eighty-six years ago.

Now local citizens, historians, and environmentalists are engaged in the second "Battle for Blair Mountain." Arch Coal and Massey Energy hold or lease land in the area near Blair. From 1980 until 1999, the union branch of Arch Coal strip-mined five thousand acres along Spruce Fork Ridge (the mountain chain that includes the town of Blair), obliterating a portion of the historic site until equipment complications halted operations. A subsidiary of Arch Coal has recently reactivated their digging here and plans to mine over 1,800 acres of the ridge; Massey holds a permit for over 1,200 acres.

In 2005, the West Virginia Archives and History Commission recommended that the National Park Service designate 1,600 acres of Blair Mountain as a preservation area. This is the fourth or fifth attempt to declare the mountain a historical site since 1980, and each time the proposal has been rejected as a result of technicalities or of lawsuits by the mining companies. Currently, the National Trust for Historic Preservation lists Blair Mountain as one of America's 11 Most Endangered Places, and locals have lobbied the state to protect the area. Amateur historian and city resident Kenny King was the first to trace the steps of the striking

Courtesy of Kenny King

Pistol discovered from Battle of Blair Mountain

miners, searching for clues along paths where his grandfather may have marched. A coal miner himself, King supports underground mining and thinks that the coal under Blair can be retrieved in a manner respectful of the historic site. In 2006, he led an archeological team headed by Harvard Ayers from Appalachian State University. Together, they documented four battle sites, collecting over a thousand artifacts in a ten-mile stretch that they believe has not been disturbed since 1921.

In the period of the Mine Wars, Blair was a sizable community. At one point, there was a motion picture house down by the creek, next to the bronze commemorative sign. Until the early nineties, Blair consisted of two hundred residences, large for a string-town occupying twenty miles along a two-lane road in mountain country. After the coal companies moved in, the population dwindled to three hundred people. In 2008, there were seventy-five.

When I visited Blair, one of the remaining residents, Melvin Cook, chatted with his cousin Ralph in his driveway, where their pickup trucks were parked side by side, content like the men to spend the afternoon under the tin overhang. They were both tall—6'2" to 6'3"—and gray-haired but athletic looking. Ralph wore shorts to his knees, socks reaching mid-shin, and a tucked-in polo shirt. His moustache sat neatly trimmed over his upper lip. A retired schoolteacher, he had left Blair for college and spent the majority of his life in another community. Melvin had slightly more stomach under a white shirt and over a pair of jeans. A coarse beard bristled just over the end of his chin. A veteran miner, he had spent thirty years of his life in the bellies of the surrounding hills.

When Arch Coal first began mining along Spruce Fork Ridge, it bought out two-thirds of the residents to prevent what companies call "nuisance suits." "Some felt lucky to get $30,000 out of a house they'd lived in all their life that retailed $10,000 in the books," Melvin told me.

After looking for homes in lower-lying communities nearer the cities, many residents of Blair found they couldn't even buy an open lot for the amount they'd cleared on their house and ended up deeply in debt. Back in Blair, in the early nineties their abandoned homes were set on fire during the night. The company never accepted responsibility for the conflagrations, although there was little doubt among the residents as to who was responsible. The message the flames sent to the remaining residents was clear. The company wanted them to get out—they were no longer welcome on their own land.

Arch Coal, in its first strip mining permit, extracted seven seams of coal, removing three hundred feet from the top of Ma and Pa Hollow, where Melvin and Ralph's grandpa used to "hunt 'coon". Explosions from the mining operation sent stray rocks into people's yards, coating everything nearby with black dust.

"You could hear the blasting like it was right there. You can set out here at like 4 o'clock and it's somethin' like night. The wife is sweeping off the porch, and then like a half an hour later she's doin' it again," Melvin remembered.

In 2008, the Arch Coal and Massey Energy companies were gearing up to resume operations on the ridge in front of Melvin's house and along the ridge behind, containing a thirty-one foot seam of some of the best low sulfur coal in the country, worth billions of dollars. Since Arch executed its first permit, 'coon hunting in Ma and Pa Hollow is no longer possible for local residents. In the 1950s, deer moved into the area, and the terrain behind Melvin's house is covered with walnut, hickory, beech, chestnut, buckeye, and white oak trees, all of which are especially good for deer. Hunting with guns is forbidden in the area, so Melvin tracks game through the woods with his bow and arrow. Some hunters have felled bucks with eight-point racks. After Arch and Massey are finished with the ridge, there will be no trees or deer left, and Melvin will be surrounded by gutted land and the few animals that can be supported by a fringe of forest on either side.

Concerning the efforts to reclaim the portion of Spruce Fork Ridge already decimated by strip mining, Ralph expressed his frustration: "They put the grass out there, but what's it good for? I call it tumbleweed. It just keeps the ground intact. It can't support any wildlife. They defend it by sayin' they got a golf course out in Mingo, but how's that supposed to make me feel better? You gotta have a rich guy to come out and play on it."

He was referring to the Twisted Gun Golf Course that was constructed over the top of the Mingo Logan Low Gap surface mine, which is touted on the website of the Mineral Information Institute as being one of reclamation's poster children: "Until this reclamation program, the counties of Mingo, McDowell and Logan did not have an 18-hole golf course due to the great expense of course development in the region's mountainous terrain." On the introduction page for the "Coal Mining Reclamation" section of the website, the Institute makes this disclaimer: "Mining inevitably disturbs land. Modern mines reclaim the surface during and after mining is completed, returning the land to useful purposes ... The reclaimed mine lands are often more attractive to wildlife and human uses than before mining started."

Similar propaganda is taught in elementary schools across coal country even as the children's families are being forced out of their homes. Mining inevitably disturbs more than just land. Before Arch Coal came into town, there was a post office and two grocery stores in Blair. Even churches—the mainstay of mountain country—are closing down. Melvin told me that the preacher that ran the Blair Baptist Church across the street discontinued services just one month earlier. Melvin would not only have to go fifteen miles to get a candy bar, but to get some preaching, too.

There used to be a convenience store next to Melvin's home. After it closed, he began cutting grass on the empty lot. When he told me this, I realized for the first time why the wide spaces between the homes in Blair were so striking. The lawns were mowed, apparently by the remaining residents of Blair.

Like many of the mountain lands of West Virginia, Spruce Fork Ridge used to belong to the people who lived there. Melvin and Ralph's "two time" great-grandfather, John R. Mullins, crossed the mountains into Boone County before the forty counties that now make up West Virginia refused to secede from the Union with Virginia and formed their own government. John Mullins once owned all the property currently being mined by the Arch and Massey coal companies and parted with the land for a nickel an acre. With his pockets stuffed with five hundred dollars, Mullins walked away feeling like a rich man. "If we woulda kept it," Melvin joked, "we'd be coal barons just like the Rockefellers."

A trail at the peak of Blair Mountain leads to the strip mining operation on the other side, an area so massive that it constitutes one of the largest permits for strip mining ever issued. Leaving Melvin and Ralph with my best regards, I set out for the road. At the first trail I saw, I parked my car on the shoulder and headed into the woods.

Patrick Thomas, 2008

Author hiking trail up Blair Mountain

The forest, like impenetrable walls of green on either side of the trail, threatened to close in with every respiring breath. Puddles of water bubbling with frogs blocked my path. Leaning into the growth, I collected several ticks on my pants. I realized what an uphill march it must have been for those old-time miners during the Battle of Blair Mountain, pushing through branches obscuring their view, stepping over the rotting logs, and dodging the machine gun fire above. Not far from the road, I was surprised to find a portion of the wall of a home collapsed in a limp pile like wet cardboard. I glanced down into the ravine along the side of the trail, where the a curious number of domestic items were splayed out—carpet slung on the side of the riverbed; a mattress down in the gulley; a couch turned upside down, with its springs in the air. Piles of smaller objects spread along both sides: a tire, a hairbrush, shoes, socks, an ice cream

Courtesy of Kenny King

Big Branch Strip Mine

carton, a coiled hose, and a child's Fisher Price keyboard. On the other side of the bank, a toilet was smashed, shards of white ceramic mixed in with the tree bark and the green leaves of ground creepers. A jar of peanut butter had settled high up in the woods. My eye caught a sheet torn from a photo album. I brushed away flaked mud from a Polaroid snapshot of a small blonde girl bending over a birthday cake blowing out her candles. A man, presumably her father, stood behind.

Emotion flushed into my blood, pushing so hard up under my ribcage that I could hardly breathe. I stepped carefully down the side of the gulley looking for clues: a pink princess lunch box, a red toy monster truck. Above the houses that line the highways in string-town Appalachia are homes tucked up into the woods, where residents lived for generations before coal companies came into town, prompting development lower down. I wondered if these were the contents of one such home, and if so, who had lived there, and why they had left so many of their belongings.

Several miles up from this gorge, the trees begin to thin, partly due to altitude, but mostly because they are a facade for what lies beyond them—a ridge that drops down into hydro-seeded pasture, rolling on, dipping, and then, after thousands of acres, climbing to another line of forest that makes a dark stripe on the top of this incredible expanse of pale green—Ralph's "tumbleweed"—the remains of what was once historic ground.

Even the visual impact of this sort of surface mining calls into question its ethics, but the makings of a home and family strewn across a ravine may have been evidence of the dire impacts of diverting those thousands of miles of streams.

"It's not a stream unless it runs 365 days a year," Ralph had said earlier, shaking his head in irony. President of West Virginia Coal Association, William Raney, once claimed that coal companies were filling in "dry hollows." The ravine below the trail was dry now, but the spoiled carpet and the couch had evidence of water damage; the objects flung on either side of the gorge indicated a significant amount of runoff had barreled through.

Amid the piles of research I've compiled for this book, I keep the photograph of the girl and her father in a gallon-sized Ziploc bag. The father is lean and mustached; his brown hair falls over his eyes as he reaches around a blond toddler to help her cut a chocolate birthday cake. In the lower left corner is another hand, pushing the cake forward. I assume this hand belongs to her mother. The mud is still caked over the piece of album page protecting the Polaroid; a dried piece of wall adheres to the back. Like my miner's lamp, this photograph is a piece of mining history that has a story to tell that I will never know. Every so often, when sorting through my papers, I find it again, and I'll pause as my heart speeds up for a moment. Fingering this photo of this child and her father, I hear the motors of two-story earth-movers still engaged in stripping away another layer of mountain. I think of the Goals Coal Processing Plant conducting business as usual in a high flood year, just waiting for an act of God to unleash the sludge lagoon above Marsh Fork Elementary, spilling disaster and heartache through the downwash of that Appalachian valley.

19

Squatter on a Gold Mine

At a fork in the mountain road up Kayford, I took a right and realized I'd gone wrong somewhere when the road ended at a fence and a guard post. I drove up and rolled down my window.

"I'm looking for Larry Gibson and a folk festival." I noticed the company name on the guardhouse and realized I was on the property of Massey Energy. I flashed a wide smile to diffuse some of the tension, and a grin crept out the corner of the guard's stiff lips. The result was a grimace.

"I don't know where it is, but you could try taking a left at the fork in the road," he conceded.

The man-made, barren lands of West Virginia can typically only be viewed and photographed from the air. Despite the fact that so many coal company executives and representatives assert this is what the good Lord intended for West Virginia, they are reluctant to let anybody view the product of their good works. These grand canyons of industrial manifest destiny are blocked from the public by company owned roads. Kayford Mountain is an exception. Here Larry Gibson holds fort against the encroachment of the Arch Coal and Massey Energy on his ancestral land and family graveyard.

Due to a number of wrong turns, I was late and the gathering was beginning to dissipate. Groups of older folks sat around picnic tables, and under a pavilion thirty people watched a country band—the audience was a combination of locals from the surrounding mountains and environmentalists. A group of young dissidents, dressed in T-shirts and sporting scraggly hair, sat on blankets on the grass. Here in West Virginia, at least MTR provides the impetus for a healthy teen counterculture.

I have been to peace protests and other sorts of demonstrations in my years of living near the nation's capital. Every time I march or hold up a sign (once a peace dove made out of a white sheet), I understand the relative futility of the gesture. I know I could be better educated in international policy, and that there may be a more effective way of expressing my discontent. I will go home, and will have little affected the world, but given myself an opportunity to vent on the National Mall in the view of the news helicopters overhead—an opportunity to say, "You suck," and "Why can't we all get along?"

In contrast, this group gathered on Kayford Mountain was a highly specialized resistance movement. Every person I talked to had a personal lean and an area of advocacy. These were not abstract objectors, but people who had witnessed the effects of MTR in their lives and in the lives of their neighbors. Appalachian Voices, the Ohio Valley Environmental Coalition, the Coal River Mountain Watch, the West Virginia Rivers Coalition—there were representatives from each of these West Virginia organizations, formed by community members and environmentalists to combat the destruction of their natural landscape and their towns. Chuck was there from Sylvester. Regina Hendrix, a member of the Sierra Club in West Virginia, was focused on garnering support for the historical preservation efforts on Blair Mountain. Regina's particular area of focus was underground sludging, an alternative to sludge lagoons, where sludge is pumped into the tunnels of old mines. Seemingly, this solution is kinder to the environment and to area residents, but in actuality it poisons the groundwater. In West Virginia, half a million people of a population of two million get their drinking water from wells, and the state's cancer rates are increasing exponentially.

Glenda Smith, a woman with shoulder length hair, dressed like a typical suburban mom, lived near me in northern Virginia. She drove out to the rally because she felt as though the coal companies were robbing the locals of their dignity. Where Massey now stripmines, her grandparents owned the land and the houses. They paid eleven dollars a month for water, but much of their food was grown on their land. They had their own fruit trees, made their own soap, and ground their own meal. Anything they couldn't make, they would walk down the road by the "Coal River" and buy at the company store. Her father worked in the mine off and on until he abandoned his wife and six children when Glenda was twelve. Glenda's mother told her that she wouldn't be able to go to college, so she took shorthand in high school. When the civil service needed secretaries, she packed her things in a cardboard suitcase and left three days after she graduated. After she got married and raised children, she wanted to show her kids where she grew up, but it no longer exists: a town called Carbon.

I spotted Larry Gibson in a fluorescent yellow shirt that read:

> We are the keepers of
> The mountains
> Love them or
> Leave them
> Just don't destroy them
> If you dare to be one too
> 1-304-542-1134
> 1-304-533-0246

This text was printed on both sides in bold black lettering. I was sure Larry had made it himself. There was no logo, no ceremony, just a message Larry was intent on blazing across the continent. I had seen him already in a number of news reports on mountaintop removal and was surprised that he was nothing over five feet tall—a small man with a mustache and a paunch. At 5'9", I towered over him. I greeted him while he was taking to Chuck.

"Maybe this isn't the time to do it because they're all juiced up. As muscular as they are, they're scared, and scared people make dangerous people." He gestured to me, "Maybe we should sic 'er on 'em."

They were discussing Chuck's cousin. Massey decided to throw a pig roast at the same time that Larry was holding his festival. The day before, Chuck's cousin had organized a group of Massey employees to beat up some of the young people—the environmentalists and the dissidents—and Larry had taken one young man to the hospital. The conflict represented West Virginia's mining dilemma: there is enormous grassroots support for outlawing mountaintop removal, but this would also destroy one of the area's only sources of bread and butter.

"They have young people here that works for the coal company on the other end of my property and they're tellin' them that I settin' on ten to twelve more years of work and I won' let them have it. They've got people here within my own family that will call the coal company when I have a meetin'. This ground is worth $450 million for the industry. Now how much do you think my life is worth now? Do you think it's worth a dang? At any time it could be snuffed away because somebody thinks they have ten years of work here."

Larry lives on Kayford Mountain. After moving down into the city, he returned when the coal companies were threatening to destroy his family's graveyard and the last of his ancestral land. "I've been doing this for twenty-one years. When I started I was as tall as this fellow here," he pointed to Chuck, "and all this has worn me down."

We cleaned up after the festival was over, and then Larry took me on a tour of what remained of his mountain, telling me his story as he walked. Larry knew that telling his story might or might not make a difference. He's told his story to CNN, to the BBC, to *National Geographic*, and to every major news station and publication in America. Just to be fair, before the interview and tour I admitted that I was unaffiliated with any news company. He told me anyway, and explained his plans to tour the country telling his story to anybody who would listen.

In 1955, the Stanley family found out that they no longer owned some of the acreage and the mineral rights of their land. Sometime between 1906 and 1909, the coal company KG Blaine bought the deed for one dollar; a transaction that the family never knew had taken place because they were illiterate. From then until the mid-fifties, the Stanley family worked land that was no longer completely theirs. Coal mines surrounded the area, so they went to Massey to talk about boundaries.

Despite the power lines running over the mountains for the coal camps, the sixty families (some with eighteen to twenty kids) that lived on the Stanley's ancestral land in the 1950s still didn't have electricity. Their only wealth was the land, and they farmed five hundred acres of it. Larry chuckled as he explained: "The major crop was moonshine. I'm not ashamed of our culture. Aside from that, honey, butter, milk, and corn." Only the company stores had canned goods, and like Glenda's folks, the Stanley family would trade their crops for clothes and tools.

Larry lives in a home that is primitive by modern standards. He pointed at a bathtub down the hill next to his garden. A trail was worn down the slope leading to it. "That's where I get my water." In 2001 he installed a cistern, but the "dynamiting blew it out." Larry has no indoor plumbing and no electricity aside from a solar panel. The winters are harsh, but Larry asserted: "We chose to be here. We haven't been displaced here. We don't need help to get off the mountains; we need help to stay on 'em. This is my people's choice for the last two to three hundred years." This was what he wanted me to write, even if I didn't write anything else.

We walked along a path in the forest until Larry stopped and pointed down. The earth had begun to gape, opening up to an intricate beehive of tunnels in the ground below. Larry estimated it went down six hundred feet: "It may go one hundred feet that way, forty feet that way, and meanders about." When he first found it, the mine crack was five inches wide. By the time of our interview, it had pulled apart to a foot and a half across. "Don't linger there," Larry cautioned. The opening was still not large enough for a person to fall through, but standing at the edge was both terrifying and mesmerizing. The light that passed through highlighted the curve of a tunnel here, an intersection there. The ground was falling beneath our feet. And no wonder.

Erin Thomas, 2007

Mine crack at Kayford Mountain

Kayford Mountain had been mined by fifteen companies over the past 130 years, and coal waste had been pumped into the abandoned mine shafts. Larry's friends gathered half a million dollars to install in a windmill at his place, but the ground is too unstable.

At the edge of Larry's property, the ground dropped off. The mountain ended, and a company sign nailed to a tree warned onlookers: private property—a last, desperate attempt to prevent me from seeing. Having grown up in a desert, I am accustomed to emptiness and distance, but not such sudden emptiness, a gouge taken out of greenness. Directly below, a field of reclaimed land sprouted pale, scrubby grass, the only foliage likely to accumulate without trees to develop topsoil. Under the thin spread of dirt, lay only bedrock. Massey had buried all the clear cut forest capable of providing fodder beneath hundreds of feet of rock and debris. Beyond the field stretched a terraced wasteland of tawny colored dirt with dark black stripes running through it. In such great emptiness, the mega-machinery used to excavate this coal appeared like miniature toy trucks in a sandbox, adding a surreal aura to the reality of such irreversible destruction.

Erin Thomas, 2007

Strip mining at Kayford Mountain

Historian and mine enthusiast Richard V. Francaviglia states that open pit mines are topological features that "appeal and repel simultaneously," and that in their development, we, within our lifetimes, can witness the sort of change of characteristic of geological time. These landscapes contain "powerful and conflicting symbolism" and do "not compare favorably with ideal landscapes because they seem unfinished, crude, imperfect, perhaps too honest a depiction of how we have treated the environment and each other." He ends by asserting: "It is in the hard places that the dirtiest work occurs to sustain our ever-demanding technology and culture."

Staring over the edge of the mountaintop removal site, I couldn't experience this landscape with Francaviglia's intellectual and historical perspective. My response came from the gut, the sort I experienced in the dark of Big Pit in Wales when I thought of five-year-old Evan. My sensibilities jerked into my chest with an almost passionate violence, and my eyes stung in anger. People don't do this, I argued. Such should not be done.

In the center of the terraces of brown earth stood an oasis, a platform of green: the Stanley family graveyard. No longer owned by the Stanley family because it was part of the land sold off a century ago, it is the high point now on Kayford Mountain. In years past, it was the low point, from which people would gaze at the mountain above. It is illegal to destroy a cemetery, so the company was mining under it. One tunnel was bored so wide, they could drive a coal truck through it. "They don' have respect for the dead," Larry said, almost spitting, "an' don' have no respect for the living."

Greg Smith, 2007

Author interviewing Larry Gibson at Kayford Mountain

"They had a sayin' out here," Larry explained. "Strong back, weak mind." In the 1920s and 30s, Massey started trading truckloads of beer for loads of coal. The theory was that if you could control them, people were like machinery. Instead of using the profits from the coal mine to promote education, the company used them to get people drunk. Larry asserts that they still do.

Standing on the edge of his forest, Larry told me his personal story, and I started to cry. It was the only response that I had for him and the gutted landscape stretched out before me. He had a family once, but his wife and daughter finally moved out, tired of the threatening messages left on the answering machine. He believes the coal company shot his dog, and that they've shot at him before. He was sure they would get him one day, too: "I'm scared. I can't tell you that I'm not concerned, but I'm more mad that they think they can intimidate people. You have to stand up in harm's way. The hardest damn thing you're going to do is convince people to stand in harm's way. They've been oppressed their whole lives and don't know what a person is. I'm out of the box. I learned as a young man that my voice counted. You know who the president was when I was five years old an' we didn't have electricity? The president was John L. Lewis, the United Mine Workers' president. We were coal miners' kids, and we knew about union activities. We knew about Blair Mountain. It was a family fight. We lost family members in that battle."

Larry has been arrested about eight times. He's been arrested at the Capitol. He's been arrested at Marsh Fork Elementary School. A few months before our interview, he was arrested again, and the policemen dislocated his shoulder. We returned back to the festival grounds, and Larry handed me a bright yellow T-shirt, identical to his own, from a box in his van. He handed me a video and then another one. I don't know that he felt any more encouraged by all his speech giving or handing out, but he did it because I came to Kayford Mountain and I was one more person that had seen mountain top removal for myself. I was a witness, and he was sending me off into the world prepared.

Larry stopped suddenly as we were walking back to my car and looked me level in the eye, even though eye contact required him to tilt his head considerably upward. "It's not so much about the fight for it. It's the fact that I'm right. If I give my rights up, they have everything else. If I give my rights up, they win. That's what keeps me going, the fact that I'm right. I don't have to convince myself. I go to bed knowin' that I'm right, I wake up confident that I'm right."

I wandered up the dirt road to the cemetery and browsed through the generations of so many Stanleys. Larry makes regular visits to the resting place of his ancestors to clean out the rocks that land there from the blasting. A ring of trees around the edge barred any view of the mountaintop removal surrounding it, and it was almost possible to believe the graveyard was set in the clearing of a grove instead of a carved edge and a sheer fall to mounds of upturned earth. With artificial flowers poking up from the ground at the foot of some graves and bows tied over others, there was a sense of remembrance. The late afternoon light filtering through the trees heightened the sense of the sacred. But it is only a matter of time before the ground cracks and the coffins fall into a maze of mining tunnels, until the earth gapes wide, and Massey will reluctantly have to give it up for lost and bulldoze the graveyard through.

Washington, D.C.

20

The Energy Future of America

As I begin to write this chapter, I become aware that I am leaning against the wall, with my back to the window, trying to make the most of the last bit of daylight. In a state like Virginia that uses mostly coal, in country that remains the world's per capita energy hog, with the whole history of coal and four generations of Welsh miners on my back, I'm in an impossible fix. Flipping a switch revs the engines of a dragline. But eventually dusk will come and then dark, and even if I neglect to turn on the light, the bright screen of my computer plugged into the wall will gleam back at me, reminding me that we are still sapping the grid.

Not far from where I sit in my home in Northern Virginia, the Greco-Roman facades of marble and local shale in downtown Washington, D.C. remind us of the past. Consistent with its stately architecture, our nation's capital has remained firmly with tradition in terms of energy. In a nation built by coal, D.C. was powered exclusively by fossil fuels instead of a more varied energy portfolio as of 2005. In a country whose history boasts some of the most revolutionary inventions of all time, our capital has yet to lead the way in energy innovation. The Potomac River Generating Station—near the borders of Virginia, Maryland, and the District of Columbia—is a relic of one of these inventions: the coal-fired power plant first demonstrated by Thomas Edison on Manhattan Island. More than one hundred years later, this plant operates on technology not much more advanced than Edison's, and it has supplied the energy margin for the District for over fifty years.

"Coal is affordable and coal is available right here in the United States of America," former President George W. Bush declared before the West Virginia Coal Association. In a parting gesture in December 2008, Bush sponsored a modification of the wording in the Clean Water Act, allowing coal companies to dump mining refuse in streambeds. Not long after President Obama stepped into the White House, the EPA asserted its authority by announcing its intentions to review some MTR permits. Immediate outcries from the coal mining industry and its supporters voiced fears that the EPA's involvement would threaten jobs and the tax base. One writer asserted that this was an attack from "environmental zealots" that ranked people "lower than vegetation." The agency's decision was lauded from other quarters as the only hope against "poisoned water, abandoned towns, and a toxic future." Although the EPA backed off after discussions with coal companies, this move at the beginning of Obama's administration had coal advocates and coal adversaries poised for mobilization on the opposites sides of the front.

Despite the fact that coal is cheap and our most plentiful energy resource, it is also our most problematic—a fact that is acknowledged across the board. According to National Resource Defense Council (NRDC) Climate Center Directors David G. Hawkins and Daniel A. Lashof, and to Robert H. Williams, a research scientist at Princeton: "coal production and conversion to useful energy [is] one of the most destructive activities on the planet." This sentiment was even reaffirmed by Spencer Abraham, the US secretary of energy during the first Bush administration: "Coal is an energy winner with one glaring drawback: it is among the most environmentally problematic of all energy resources."

Coal is energy in one of its most impure forms. This rock of ancient marine and tree decay is mixed with sulfur, which, when released into the atmosphere, returns to the earth as acid rain. Coal often contains traces of poisonous metals like arsenic, fluorine, and selenium. In developing countries, these metals are causing significant health problems for individuals that burn coal in household stoves and fires. Coal also contains radioactive chemicals, such as uranium and thorium, and more than thirty years ago it was discovered that radioactive poisoning from coal burning plants has a more significant impact on the surrounding populace than routine radioactive emanations from nuclear power plants.

The Potomac River Generating Station is no more than seven miles from my home in Alexandria, Virginia. The technology in this plant is antiquated, and the output for much of its recent operation is illegal by current emission standards. Five smokestacks stand tall over the Potomac, still sending soot into the air. Our nation's capital must be fueled, even at the cost of ignoring its own legislation.

This is not the only grandfathered power plant in Virginia. Eight out of the ten coal burning plants in the state are outdated, pouring thousands of tons of nitrogen oxide and sulfur dioxide into the air. These oxides cause ozone smog, which descends heavily on Northern Virginia during the summer months, aggravating lungs, including my own. On these inversion days, I feel the pollution before I smell it or see it. The air takes on almost a granular quality. There is a lightness in my breath, and I know if I go running I will come home early with my head pounding and the taste of blood in my mouth.

Asthmatics are not the only victims of coal pollution. Many of Virginia's streams have become too acidic to support trout, and the greatest threat to the Chesapeake Bay is nitrogen, a substantial amount of which enters the water from the air. Virginia farmers lose millions of dollars of cotton, soy, wheat, barley, and peanut crops each year to ozone smog.

If the immediate health impacts do not present a sufficient case, there is also the fact that coal burning is the leading contributor to global warming. If the eight grandfathered coal plants in Virginia were brought up to current environmental standards, it would be equivalent to taking four million cars off the road. Coal fired power plants are the leading producers of carbon dioxide (CO_2), and the United States tops the world in CO_2 emissions, producing 25 percent of the global total for 4 percent the world's population. 2000 to 2009 was the hottest decade. The long-reaching effects of coal have the potential of bringing on environmental disasters more destructive to human life than the whole history of coal mining explosions.

Under the Obama administration, hot debate rages in Washington, D.C., between agencies with competing ideologies that lobby for the energy future of America. Some fear a "carbon-constricted future," and others the "coal rush"—the rapid increase in coal burning plants being built or on the drawing table. Policies are made on the city and state level concerning the fate of coal, but with our country on the brink of an energy crisis, considerable power lies within the national realm to make decisions with far reaching consequences. Civilizations have risen or fallen based on their use of resources.

The United Mine Workers of America is among the organizations that send advocates to the national legislature to lobby for coal. This is part of its primary purpose: to preserve the jobs, benefits, and rights of the coal miners and other workers it represents. Despite the corruption during the John L. Lewis and Tony Boyle administrations, the UMWA has played a significant role in the history of coal mining. Founded in 1890 in a merger between the Knights of Labor and the National Progressive Union of Miners, the UMWA pushed through the eight-hour workday in 1898. They are responsible for the many of the major mining clashes throughout the twentieth century: the Latimer Massacre, the Battle of Virden, the Ludlow Massacre, the Battle of Blair Mountain, the Herrin Massacre, the Besco Strike, and the Harlan County War. UMWA rabble-rousers such as Demolli and Mother Jones pepper the history of my mining ancestors.

In its heyday, during the reign of Lewis in 1920, the union represented half a million miners; the UMWA currently speaks for two hundred thousand—30 percent of American miners.

The UMWA's D.C. office is located in Northern Virginia near my home. The decor in the lobby and halls was Spartan; nothing but large black and white photographs of miners hung on the walls, giving the office the feel of a coal miner's hall of fame, memorializing man's epic encounter with the earth in search of energy. In the conference room hung almost Chuck Close-sized portraits of union leaders: John L. Lewis and the current president, Cecil E. Roberts.

Phil Smith filled a leather upholstered chair with his broad shoulders, projecting a brawny but articulate competence. He is the communications director at the UMWA and looked the part of a miner made union leader, but he had spent all of his working life behind a desk. His father was an organizer in a steel union in Texas and inspired Smith to become a "student of labor history." When I asked Smith about the energy future of coal, he leaned back in his chair confidently: "We believe coal has a strong future. Right now, 53 percent of our energy is generated from coal. If you set aside the improvements in technology that we anticipate in the next fifteen to twenty years, coal would still play a large role in generating electricity for the next twenty to twenty-five years. Right now, nothing can replace that, and that's just a technological fact. There aren't enough wind farms and there aren't enough solar farms, nuclear power, hydroelectric power, however you want to put it. So we're going to continue burning coal, but the good news is that as we move forward, the development of carbon capture sequestration technology is going to play a huge role, maintaining coal as a key base of our energy. If that technology can be developed and implemented, there's another hundred years of easily recoverable coal. There's another hundred years after that of coal that's going to be a little bit harder to get at."

Wind and solar energy are gaining greater portions of the market share, but even if the technology is there, the implementation is not. Smith is correct: coal is going to continue to be America's leading energy source for some years to come as long as there is the political will to burn it. Our society is reliant on coal; it's a security we are reluctant to give up. In a recent Senate hearing concerning carbon capture sequestration, Senator Byron L. Dorgan confirmed this sentiment: "We're seeing a lot of new things happening in the renewable fuels area. But this should not suggest that we are not going to use our fossil fuels. We are. I don't think anyone believes that somehow in a short, intermediate, or long term that we're not going to continue to use fossil fuels. The question isn't 'whether.' The question is under what conditions we use them."

According to Smith, carbon capture sequestration (removing and storing the carbon dioxide from gases released by burning fossil fuel, thus preventing it from reaching the atmosphere) would cost a billion dollars a year to develop and commercialize. Carl O. Baur, Director of the National Energy Technology Laboratory, testified before the Senate Committee on Energy and Natural Resources that the "economics are prohibitively expensive." If the carbon dioxide emissions were captured from just half of the coal fired power plants in the United States, on average our electricity costs would rise from twenty-five to eighty dollars a megawatt.

Under the restrictions of a carbon constrained world and possible carbon cap and trade legislation, carbon-capture technologies are the coal industry's best bet, and interested parties have been quick to adopt "clean coal" rhetoric in promoting its implementation. The American Coal Council (ACC) is among the organizations at the forefront of the D.C. political debate. Janet Gellici, its CEO, has been with the council since its onset. Middle-aged, with gray hair down to her shoulders, her history in coal is similar to mine: her great-grandfather mined coal in Pennsylvania. In addition to her responsibilities at the ACC, she serves on the National Coal Council (a governmental advisory board) and on the board of directors for the Women's Mining Coalition. Although the ACC sends representatives to congressional hearings on energy, according to Gellici, it is not a lobbying group: "We deal with the full spectrum of the coal chain from extraction to the end consumer, to energy traders, and coal support services. The American Coal Council is not a lobbying organization, but advances the business interests of coal."

The ACC represents over 160 coal traders, coal suppliers, coal transportation companies, coal support service firms, and coal consumers in North and South America, and one of the goals listed on the ACC's website is promoting coal as "an economic, abundant/secure and environmentally sound energy fuel source." *Environmental* and *coal* are two words the carbon industry has been eager to tag team. This is a perspective Gellici was also eager to propagate: "The industry is making strides towards clean coal technology. There's a lot of hope there, but it depends on the political will for carbon capture. Energy generation needs to be done cleanly. I don't think anybody in the industry would disagree with that."

As an example, Gellici mentioned a plant in Beulah, North Dakota. She was speaking of the Great Plains Synfuels Plant, which manufactures natural gas from lignite coal, a loosely bound solid. Lignite is the lowest grade of coal, containing only 25 to 35 percent carbon. Bituminous, burned by most plants in the United States, is a soft coal that consists of many impurities mixed with its 60 to 80 percent carbon. Anthracite is the rarest and hardest. Made of at least 90 percent carbon, anthracite is difficult to ignite and burns with a low blue flame. Most plants would not deem extracting energy from lignite worth the effort, but the Great Plains Synfuels Plant began operation in 1984 as a collaborative project between several energy companies and the Department of Energy (DOE) and is currently the cleanest energy plant in North Dakota, capturing two-thirds of its CO_2. This CO_2 then travels through pipes to the oil fields of Saskatchewan, Canada, where it is used in oil recovery projects.

According to Gellici, there are several other options for generating coal-powered energy besides traditional pulverized coal plants, which transmit only 30 to 40 percent of the energy contained in the coal to the power lines. New Supercritical and Ultra Supercritical coal burning plants can achieve 10 to 20 percent more efficiency by increasing the heat and the pressure at which the coal is burned. Intensified heat and pressure produce a higher quality steam in a *supercritical* state, where the difference between a gas and liquid is indeterminable. There are four hundred Supercritical and Ultra Supercritical coal plants around the world; despite China's ascent to the world's top CO_2 polluter, the Chinese build one of these plants every month. The United States has been much slower to retire old plants and take advantage of this new technology. Almost all of our coal based energy comes from "technological dinosaurs built in the fifties, sixties, and seventies."

An integrated gasification combined cycle (IGCC), which uses chemical reactions to separate the constituent parts of liquefied coal, is an even more advanced and much cleaner technology. Gellici continued: "There is the Tampa IGCC plant in Polk, Florida, and the Wabash River plant in Indiana. But for these new plants, public utility commissions have being denying permits. They can't justify new power plants because it's not the least cost option. When they do get the permits, IGCC plants get shut down by the environmental community because of their objection to building any new coal plants."

The process used by the Tampa and the Wabash River IGCC plants involves pumping coal mixed with water into a superheated oxygen chamber that separates the coal into its constituent parts. The resulting liquids and gases are then put through filtering processes to remove sulfur, particulate matter, and nitrates. Next, the syngas is ignited to turn the turbine—the first cycle in a combined-cycle system. Finally, the exhaust gas is used to heat water into steam, which turns a second turbine. Burning the liquid at such high temperatures consolidates the CO_2, which makes it easy to capture. However, because carbon cap and trade systems have yet to be put in place, these IGCC plants pump easily capturable CO2 into the atmosphere.

These two plants were constructed as pilot facilities for IGCC technology and subsidized by the federal government. The Polk Power Station has been running commercially since 1996, and the Wabash River Station has been producing electricity since 1995, a wise investment since the older broilers at the power plant were ordered to be shut down in May 2009 due to violations of the Clean Air Act. Tampa Electric, the company that owns the Polk plant, has been able to derive profit not only from the electricity, which due to 15 percent more efficiency is cheaper to generate, but also from the captured pollutants by selling sulfur to fertilizer companies and slag to the cement industry. As of 2007, 159 new coal-fired plants are in the planning process, but only nine of these will employ IGCC technology because these plants cost one billion dollars apiece to build, which is 15 to 20 percent higher than the cost of a traditional coal burning plant. In order to encourage companies to invest in IGCC and carbon capture sequestration (CCS), the DOE is planning to construct FutureGen, a plant that will combine both of these technologies.

Although the coal industry is quick to adopt "clean coal" rhetoric, the few utility companies willing to actually invest in cleaner technologies is evidence of the industry's tendency to cling to old, destructive practices until forced to comply by federal legislation. CCS, for example, is much easier to implement at an IGCC plant than at a traditional pulverized coal station. Choosing to stick with tradition, in this case, is a bad investment in the future on more than one front. Even if not initiated by the Obama administration, a legislated carbon cap is still inevitable, which will make pumping CO_2 into the atmosphere a costly business. Likely, the energy companies building traditional coal burning plants plan to look for loopholes as they did when the government passed the Clean Air Act, thus allowing so many grandfathered plants (like the Potomac River Generating Station) to stay in operation. But as *Washington Post* columnist Robert Samuelson asserts: "the lifestyles that produce greenhouse gases are deeply ingrained in modern economies and societies." All the coal industry has to do is threaten a blackout, and critics back off.

Throughout our interview, Gellici had maintained a comfortable and easy manner. Her voice contained no element of defensiveness, despite the fact that she was representing an industry facing significant criticism. When mentioning the possibility of an energy crisis, she transitioned amiably: "The NERC [North American Electric Reliability Corporation] did an electrical assessment, and in 2009, some states won't have enough of an energy margin, falling short by 8 to 9 percent. In the future, there will be more plasma screens, more hybrids, more plug-in vehicles. In Italy, citizens have marched out on the streets because of their utility bills. Germany is part of the Kyoto agreement, but can't keep its commitment and is building new coal plants. China surpassed us in emissions last year. CO_2 emission reductions need to be a global commitment."

Gellici ended with a statement that embodies the energy conundrum. "It's like the sign in my car mechanic's shop: 'We can do this cheaply, quickly, or right. Choose two.'" In order to prevent global calamity, "quickly" may not be negotiable. According to John Holdren, professor of environmental policy at Harvard University: "If all the coal-burning power plants that are scheduled to be built over the next twenty-five years are built, the lifetime carbon dioxide emissions from those power plants will equal all the emissions from coal burning in all of human history to date." China erects approximately one power plant a week and is expected to construct 562 new plants between 2006 and 2014.

In the lobby of the National Resource Defense Council (NRDC) office of Washington D.C. hangs a poster: "Some mistakes are no big deal. Global warming isn't one of them." According to its website, the NRDC is the "nation's most effective environmental action organization." Founded in 1970 by law students and attorneys, the NRDC employs the expertise of three hundred science, law, and policy experts and the grassroots support of 1.2 million members to protect the natural environment and the creatures that live within in it. They are at the forefront of the energy debate, attempting to combat global warming and help America move beyond fossil fuels. Jim Presswood, the agency's federal energy policy specialist, grew up in Knoxville, Tennessee. He spent his summers at his grandparents' cabin in the Smoky Mountains, where he developed the love of nature that finally propelled him into the national arena. Initially a public policy major, his particular interest in air quality began as a college senior, when he waited at a traffic light behind an old pickup truck that spewed thick smoke out its exhaust pipe. Accordingly, Presswood holds a very different view on coal than Janet Gellici: "We do not support additional coal generation. Other forms of energy are sufficient to meet our resource needs, but if coal needs to be deployed, we think that it should be IGCC with carbon sequestration technology disposal. People kind of think of the NRDC as pro-coal, but we're not. We recognize the political realities of China going gangbusters to develop its coal industry. Our coal lobby has a really strong influence in Congress, and we recognize that there's going to be a least a $100 million budget for carbon capture plants to be built. We tacitly oppose conventional pulverized coal technology, no questions asked. They do not belong in a carbon constrained world where we have to meet increasingly aggressive reduction targets. We think ultimately efficiency and renewables will win out. They'll be able to beat the socks off coal on a cost basis. If you internalize carbon costs, you level the playing field like it should be. It's all about internalizing externalities."

When Presswood mentioned internalizing coal costs, he meant the cost coal has on our society that is currently being absorbed by taxpayers. This includes the health care expense of black lung, which claims one thousand miners a year, and air pollution, which claims 2,400 people, and then the medical bills of those living near coal mines and cleaning facilities that are poisoned by their operations. Added to this is the price of injury and death incurred by miners and the tax dollars supporting the budget of the Mine Health and Safety Association, an agency whose sole responsibility is to prevent such accidents. A third element would be the amount required to reclaim the 12,204 abandoned mines documented by the Bureau of Land Management and clean up the damage created by flooding and coal sludge spillage in areas such as Buffalo Creek, Pittston, and the Tennessee Valley Authority. Finally, this would include the cost of cleaning all the waterways in America that have been polluted by mining—all in all—a cost almost innumerable.

Presswood's opinion reflects the consensus among many environmentalists. James Hansen, director of the National Aeronautics and Space Administration's Goddard Institute for Space Studies and a leading climatologist, asserted before a crowd in D.C.: "Until we have that clean-coal power plant, we should not be building them. It is clear as a bell." He summed up his discourse with the radical suggestion that by 2050 any "dinosaur" plants that aren't capturing CO_2 should be bulldozed. According to climate scientists, safe CO_2 levels that would help prevent violent climate changes should be somewhere under 450 parts per billion. Maintaining this level would not only require quick and efficient adoption of clean coal technologies, but also improvements in energy efficiency and renewable generation.

Any action that is effective and rapid is going to cost money. When I asked Presswood which technologies he felt were worth the investment, he indicated that he favored a varied portfolio: "In order to achieve reduction of global warming pollution in a way that is cost effective, we really need to incentivize energy technology. We need to have the incentives structured in a way that they're based upon performance. You don't want to pick technologies; let the best technology win."

Wind is the fastest growing alternative energy source and presently constitutes 5 percent of the energy grid. Prices for wind generated electricity are already competitive with natural gas and are likely to continue to decrease. Wind's primary drawback is its intermittency. Although turbines can't provide a steady supply to meet demand as easily as coal fired power plants, recent storage technologies enable wind power to be a more consistent source of energy. Night is the generation peak for wind plants, and energy not needed for the grid is used to heat liquid, which can be used later to heat water to generate steam. This way, during periods of lower generation, wind energy can be supplemented with steam power. It is anticipated that wind may be able to carry 20 percent of the nation's power needs by 2020 to 2040.

In 2009, solar energy powered .5 percent of the grid, and many are hopeful its price will be competitive with natural gas and wind in five to ten years. According to some scientists, concentrated solar power (CSP) systems can be developed to carry a more significant portion of the power grid. Charles E. Andraka of Sandia National Laboratories asserted before the Senate in 2006: "The Southwest US has an incredible resource. We like to call this 'the Saudi Arabia of solar energy.'" To date, CSP researchers have identified easily reachable sources of seven terawatts of generating capacity, which is seven times the current electricity-generating capacity of the United States. CSP is a fairly simple technology, using mirrors to concentrate sunlight and use its heat to drive turbine engines. This process is incredibly efficient, with 31 percent of the collected solar energy being converted into power. It also has some advantages over wind power. It's more easily stored and doesn't require batteries, enabling it to meet peak demand periods on the grid. Potential disadvantages to CSP include the initial costs of building the plants, property taxes associated with the use of the land, and the length of the transmission lines. Most of our solar resources are located in the desert, far from energy users and cities.

Like Gellici, Presswood foresees an electrification of the transportation industry and an increase in hybrid vehicles. Where Gellici would advocate syngas, Presswood supports the extended use of biofuels. In the United States, ethanol is produced primarily from corn grown in the Midwest. Currently all vehicles can run on gases that contain 10 percent ethanol, and vehicles that run on diesel can accommodate up to 85 percent ethanol. Concerns over ethanol deal with the amount of energy that is used to grow the biomass and subsequently convert it into a source of power. Ethanol also contains less energy per mile than oil, which raises fuel costs for consumers. From environmental and geopolitical standpoints, ethanol reduces America's reliance on foreign energy sources, replacing them with a gas that burns cleanly and is renewable. From an economic standpoint, ethanol also brings jobs to rural communities.

Among the technologies Presswood champions is the often ignored energy efficiency: "Energy conservation is flipping on and off the lights, but efficiency is getting the same amount of work, but using less energy. Buildings can reduce energy consumption by 30 percent, and that adds up to a lot of emission reductions." The idea of energy efficiency was first tossed around in the early seventies after the 1973 oil crisis, but it hasn't been until recently that governments, businesses, and individuals have looked into this largely untapped energy resource.

Energy Star is a project founded jointly by the EPA and DOE to help Americans conserve both energy and money. In 1992, the Energy Star label was introduced on computers, and now the label is available on home electronics, major appliances, office equipment, lighting, homes, and commercial buildings, certifying energy efficiency. In 2008, Energy Star helped Americans avoid the equivalent of greenhouse gas emissions from twenty-nine million cars and save nineteen million dollars on utility bills. Amory Lovins, the Rocky Mountain Institute's cofounder, chairman, and chief scientist, is an efficiency guru who has designed energy saving alterations for cars, trains, factories, and housing. He believes half of current oil use can be offset with efficiency, which he defines as "wringing more work out of our kilowatt/hour by substituting technology and brains for fuel and money." He also notes that "efficiency is arguably the highest return/lowest risk investment in the whole economy." It has been asserted that up to 75 percent of US electricity needs could be saved with innovations to enhance efficiency, which could also keep growth in the global demand for energy at less than 1 percent per year.

According to Presswood, legislation plays a strong role in implementing these new technologies. "Having an energy resource standard would be key. Renewable production standards goals would be key. But of course the first, fourth, fifth priority would be getting a cap and trade system in place. If you have the right price in the wrong market, it's going to be really difficult to get any of these technologies up and rolling. Our goal is to reduce carbon emissions going into the atmosphere, which is just something that we have to get under control."

Voices like Phil Smith's, Janet Gellici's, and Jim Presswood's collide in a national debate that will ultimately decide the energy future of America, but I wanted to get an opinion from the scientific sector—the researchers that add fuel to the fire. Greg Jackson is the acting director of the University of the Maryland's Energy Research Center and has conducted extensive research into fuel cell technology. I interviewed him in the top story of the Engineering Department, a floor that perches like at attic over the rest of the building, accessible by climbing four staircases. He was thin, balding, and birdlike, nesting in an office that simultaneously suggested "science guy" and "environmentalist."

Like Gellici and Presswood, Jackson anticipates a future where the electricity and transportation industries converge. His possible solution for *stationary power* is one that has been shunned for most of recent history by the environmental community: "We have the potential to increase our nuclear capacity, not quickly, but we have the resources to do it, and maybe not the political will, but with global warming becoming more and more important that political will is going to slowly show. The further we get from Three Mile Island and the misrepresentation of what happened there, society will find it more acceptable. The question with nuclear power is how we deal with the waste. Nuclear is good, but it's clearly not the catch-all answer."

In 2009, nuclear fission powered 19 percent of the grid. A technology pioneered in the 1950s, nuclear power was considered the fuel of the future until fossil fuel costs went down in 1970 and concerns rose over the disposal of processed uranium. Building a nuclear power plant is significantly more expensive than building a coal fired plant, requiring utility companies go heavily into debt with the hope of recouping their investment from future electricity generation. The switch back to coal intensified when the construction of new nuclear power plants came to a complete standstill in 1979 after the partial meltdown of one of the reactors at the Three Mile Island plant. Little toxic material escaped, and subsequent epidemiological studies have confirmed that the incident had no impact on local incidence of cancer, but the public hysteria created by the accident sealed nuclear power's unpopularity in the United States. But, as Greg mentioned, in the wake of global warming concerns, nuclear power has a shot at comeback.

Jim Presswood's point of view on nuclear reflects society's hesitation: "As soon as they take care of storage costs and the environmental issues associated with storage, the operation risks, and the economics, yes, we're open to nuclear. We just haven't seen technology yet that avoids all those very real problems. Each technology carries risks, but the risk of nuclear technology is relatively high." Patrick Moore, an environmentalist who helped Greenpeace in the 1970s, disagrees. He feels that Three Mile Island, the sole—and nonfatal—accident in the history of US nuclear power generation, was a success story because the concrete containment bulwark prevented the radiation from escaping into the surrounding environment. According to Moore, "Nuclear energy is the only large-scale, cost-effective energy source that can reduce ... emissions while continuing to satisfy a growing demand for power ... Today there are 103 nuclear reactors quietly delivering just 20 percent of America's electricity. Eighty percent of the people living within ten miles of these plants approve of them. Although I don't live near a nuclear plant, I am now squarely in their camp." Uranium still contains about 95 percent of its energy after the first cycle, and it can be reprocessed, reducing wastes. Japan, France, Britain, and Russia all recycle uranium, and this technology is currently being considered in the United States.

But nuclear, according to Jackson, would not be the ultimate destiny of energy: "I think the end goal is clearly to move to sustainable energy production with solar and wind, and at least the hydroelectric that we have. The challenge is that takes up a lot of land, and it's not environmentally perfect because there's loss of agriculture and loss of forests, and you don't want to do that. In other parts of the world they're chopping down forests, so they can grow biomass for making ethanol. It makes sense for profit—it doesn't make sense for global warming."

Jackson continued by explaining how the power-generating capabilities of wind and solar are yet to be seen, and even these have their problems: "Arnold Schwarzenegger is pushing for a big solar farm out in the Mojave Desert, and environmentalists are against it, and understandably because they're worried about habitat. They have their point. It's ironic, though, that the people that are against installing solar are the environmentalists." To this point in our interview, Jackson's responses had reflected the opinions common to many environmentalists. His next argument revealed the scientist and innovator: "One thing that isn't getting a lot of press right now, but is certainly possible, is if we boost our electricity grid, you can actually make fuels by using electricity to split water and carbon dioxide back into fuels. If you're making your energy cleanly, nuclear and renewable, and you have excess, you can use some of that excess to make hydrogen cells. The thing about renewable energy is it's not consistent. If you rely on it, you want to make all the power when you can. How do you store it? Batteries are environmentally disastrous, too. The only way to store it is to use the excess to make fuel, so you're storing your excess energy in chemical bonds, which you'll use for airplanes or trucks, and for cars if we are still using fuel-based cars. If we could get to a hydrogen-based infrastructure, it would be wonderful."

Hydrogen fuel cell technology extracts the energy stored in the cellular bonds of hydrogen, separating the protons and electrons. These atomic particles then combine with oxygen to produce the only byproduct of the process, water. The main problem with hydrogen fuel cells to date is cost.

When I asked Jackson which of all energies is worth investing in, he mentioned geothermal. This will happen "when it's economically viable," he believes. "It's expensive to start, but once you get it going, it's pretty easy to operate. Drilling these small holes deep into the earth is very capital intensive. The reason that nobody builds them is nobody wants to put up the investment."

Geothermal is a little talked about technology that has the potential of carrying a significant portion of the power grid. Under the earth's surface lie areas with elevated temperatures near enough to the crust that the heat generated by magma or plate tectonics can be harnessed to heat water and create steam. This steam, consistent with conventional energy-generation processes, turns turbines. As of 2010, there were nine states with geothermal power plants—Alaska, California, Hawaii, Idaho, New Mexico, Nevada, Oregon, Utah and Wyoming—and several others have known geothermal resources. The estimated number of undiscovered geothermal resources is more than four times those currently identified in the United States. There is also the possibility of drilling in permeable rock to reach heat sufficient to generate 517,800 megawatts of electricity, which, at current usage, is enough to power half of the electrical grid. The cost of drilling these holes can be prohibitive, but on the whole geothermal is much more cost effective than IGCC plants and carbon capture technology.

Jackson, like everybody before him, had something to say about coal: "We have a lot of coal, and there's too much industry invested in coal power for that to just go away as much as, personally, I would like it to go away. I just don't like the way coal mining is destructive, and particularly the way it's done in West Virginia. I'm not thrilled about carbon sequestration because it relies too much on very good execution and accountability. I have colleagues who are adamantly opposed to it because they are afraid that the carbon dioxide will burp. I actually think you can minimize that risk. Carbon dioxide is a heavy gas. It likes to stick to minerals. I think you could easily store it underground."

The stock example for carbon burping is Lake Nyos in Cameroon, where a sudden release of volcanic CO_2 in 1986 asphyxiated 1,700 villagers as well as their livestock, a scenario that carbon sequestration advocates consider highly unlikely to occur again.

No matter what energy sources we invest in, Jackson asserts: "Conservation has to be a part of the equation. You can't force people with conservation. I work in a center for environmental energy engineering, and I see regularly a lack of will to recycle among the students, and we talk about these things. You ask yourself, if we can't get them to do, who can we get to do it? I think the only way is to tie it into money. We have to pay for energy. That's when society seems to move."

As a scientist, Jackson spoke mostly of solutions, but he ended with a statement that left me slightly chilled: "The transition, whatever we do, will be long, and it makes me very nervous. I agree with the International Panel on Climate Control that we're at a dangerous point. If we don't take action soon, from a climate perspective, we'll be at a point where we just can't adapt. We're not going to stop the problem, and maybe we're there already, and unfortunately this transition is going to be long, and that makes me nervous for my kids. Once you stress a world like that, injustices get amplified."

21

A Drop in the Bucket

The red brick exterior of the Potomac River Generating Station squats on the banks of its namesake, a body of water that has provided an easy outlet for heat and toxic chemical disposal over the sixty years of its operation. Perhaps most significant about its appearance are the five stubby smokestacks, built at a height as to not interfere with pilots landing at Reagan Airport across the river. Tracks are laid out alongside the plant, and a train with a blue engine delivers the coal. The Mount Vernon Trail runs in front of the generating station, and several joggers passed by as I approached the entrance to take a tour of the facility.

The event started with breakfast laid out in a break room—donuts, bagels, and fruit platters—providing an opportunity to mingle with the other attendees and plant employees. A woman appearing to be in early stages of retirement sat down across from me. Her gray hair curled around the outside of her face in the perma-hold of elderly women's hair. She wore a thick silver-beaded necklace and a flowered jacket. "Did you know that recycling one can powers a TV for three hours?"

A woman in a denim jacket next to me responded, "Taking care of the environment when I was growing up meant you sunk your cans."

I must have looked at her quizzically.

"You know, fill them with water and sink 'em instead of letting them float around."

The prim woman across from me shifted the topic with another environmental statistic, seemingly off the top of her head, "Plastics kill one million sea creatures every year."

"What kind of creatures?" asked the woman in denim.

"Well, sea creatures."

I learned later that these were statistics printed out in a display across

Erin Thomas, 2011

Potomac River generating station

from the food, indicative of the plant's environmental concerns.

Somebody down the row nodded, "Yeah, plastic bags can get stuck around ducks' necks."

"So why are you here?" the prim woman asked the one in denim.

"My son works here. He has a degree in engineering."

"Green engineering?"

"A degree in engineering, been graduated a year. His dad came down already. I decided I would come see what it was all about. He loves it here." Then, with a sly look, the woman in denim asked the prim woman, "Are you a city resident?"

"Of Alexandria? Well, yes."

"Are you for or against the plant?"

"Well," the prim woman paused, choosing her words carefully, so as not to reveal too much. "That is what I am here to see."

"The Mirant Potomac River Power Station burns low sulfur coal from central Appalachia, which includes southwestern Virginia and Kentucky," began Mike Stumpf.

He meant this as a plug for the plant, but somebody in the crowd had been watching the news.

"Isn't that where they cut down all those mountains?"

"No, all the mines in central Appalachia are deep mines. Mountaintop removal takes place in the northern part of West Virginia. If we don't agree with how our suppliers get their coal, we don't do business with them," responded Mike with a very moral look. He had a long face, and when he straightened his lips, it stretched even longer.

The coal ribboned through the Appalachians is low sulfur, clean burning coal. In 1990, with the passage of the Clean Air Act, demand for low sulfur coal increased dramatically as power plants endeavored to comply. As the air became cleaner, mountaintop removal became the leading method of coal extraction in West Virginia. Kayford Mountain, Marsh Fork, Sylvester, and Inez are all in central Appalachia.

Mike was presenting to what, in general, was a nay-saying crowd. The majority were residents from the nearby City of Alexandria, which, at the time, was deadlocked in a lawsuit with the Potomac River Generating Station for polluting the air, polluting the sound, and being a neighborhood eyesore. This strip of Northern Virginia where the plant is located used to be an industrial suburb of Washington, D.C., after World War II. From a bird's-eye view, the plant's red exterior complements the green of the trees that surround the buildings in Alexandria. It would make a pretty picture for industrial era nostalgics, the power station a remnant of former times. From ground level, it is a surprising landmark so close to neighborhoods and the highrise of Marina Towers that equals the smokestacks in height.

The prim woman from Alexandria raised her hand, "Isn't it true that energy produced by this plant doesn't benefit Alexandria residents?"

At this point, Misty Allen, a lawyer hired by the plant to promote their public image, chimed in: "That isn't exactly true. All energy enters the grid. It's all connected. The Pepco [Potomac Electric Power Company] grid is right out front. Energy is generated exactly to meet the demand. The electrical pressure is always the same, sixty cycles per second." To be specific, the Potomac River Generating Station produces 482 megawatts, servicing 482,000 homes in the PGM energy market. But the Potomac River station does not, in fact, benefit Alexandria residents. The power it produces is siphoned off into energy lines that head north toward the District of Colombia and Maryland. The City of Alexandria is serviced by the Dominion power grid.

My breakfast fellow's question got the crowd going. Another visitor took courage, "Does the plant pollute the river?"

Mike fielded this one: "The wastewater is filtered before it leaves the plant, and the pH adjusted. It only is fifteen degrees hotter, but in the Potomac it's like a drop in a bucket."

Heat pollution always seems like a minimal issue, but it can alter the ecosystem of a body of water. The Geneva Steel factory that my great-grandfather Zephaniah applied to after leaving Castle Gate was still in operation until 2001. A welcome development in World War II times, it provided jobs for many residents more mechanically inclined than Zeph. Over the years, the heat the steel plant released into Utah Lake caused an overabundance of algae. Most of the native species of fish have died off, overrun by the sturdy Asian carp that Americans don't have the patience to cook long enough to eat. Utah Lake is known for being rough due to the mountain winds, and many swimmers have died in its waters; from under the murkiness and overgrowth, their bodies have never been recovered. As a teenager waterskiing on Utah Lake, I was always afraid of what would surface.

"Maybe you could dump ice cubes in the river every hour, you know, to even it out," offered a man with a long white braid down his back.

We file out of the presentation room and put on hardhats, goggles, and earplugs that prove to be essential. It is my first tour of a power plant, having only walked through buildings built in the seventies and early eighties that were designed to resemble power plants. This was before the environmentalist movement took full swing; industrial was in. Utah Valley State College, where my dad worked before he started at the College of Eastern Utah, looked like a power plant. The Pompidou Center in Paris is meant to imitate a power plant. The Tate Gallery in London actually was a power plant. They are all strikingly accurate. In the plant I was touring, painted pipes of different diameters crawled the cement walls; everywhere were valves and wheels. The hum of the equipment was deafening.

The Potomac River Station has five generators, but only three were currently operating, a result of the lawsuit between the city and the plant. The production of energy follows the same basic principles of the first steam engine—designed by Thomas Newcomen in 1717 to pump water out of a coal mine—except instead of moving a piston, the steam turns a turbine. Coal is conveyed into a pulverizer, and then the dust is pumped into the broiler where it ignites into a ball of flame. The heat evaporates water into steam, which turns a turbine, which spins a magnet in a generator. Coiled around the magnet is a long wire that catches the electrical current from the magnet, converting steam power into electrical power. This specific technology hasn't changed for over seventy years, but the operation requires constant watch: the most common problems are caused by wet coal, rocks, and fan malfunction. After the coal dust is consumed, the ash falls to the bottom of the broiler. After a cycle, the steam is cooled, and then the water is filtered to be pumped back through the system in a continuous loop.

Lynwood Reid, the operations team manager, led our group. He had a winning smile, so winning that Mirant posed Lynwood cultivating their wildlife garden for one of their brochures. A large black man, he worked for Mirant for thirty-seven years, entering the plant just after high school.

His job for the day in his words was to show us, "What's our intentions and how we doin'."

After taking us through the generator rooms, we filed into the control rooms, where he cautioned us not to lean against the gauge- and knob-covered panels of the old supercomputers. Here we met Larry Lane and a handful of other employees who managed their modern-day counterparts of the retired equipment, watching fireballs that flickered across the TV screens in green-orange and amber. It was a Saturday and the plant was operating on low, but on regular days, 133 employees worked at the plant. On average, they made sixty-five thousand dollars a year, were forty-seven years old, and had worked at the plant for twenty-one years. I could understand why none of them were too eager to back down to the City of Alexandria's complaints without a fight.

The roof, reached by climbing up a steel-grated staircase, was our last stop in the plant. Throughout the tour, the group had strained to hear Lynwood through their earplugs and over the hum of the equipment. On the roof, it was silent except for a beep. The blonde middle-aged couple and their adopted son, the young Asian American and his tow-headed wife, and the two female yuppie roommates all discovered that they lived near each other in Alexandria.

I looked out over the Potomac River and the trees, businesses, and apartment buildings that lined its banks. The coal was piled and terraced in a long field next to the plant, and the yellow scoop had paused in its operations until Monday. At the edge of the coal piles stood a fence, then trees, then Marina Towers, one of the plant's chief adversaries and the recipients of the plant's downwash during bad weather conditions. The intricate wire web of the Pepco grid buzzed out front.

"What's that noise?" one of the yuppies asked about the beep. "I hear it when I wake up at night."

"Yeah, we do, too," the young couple claimed.

Lynwood looked nervous. He put his hands in his pockets and hemmed and hawed. "I dunno actually. I'll 'av to ask Mike."

The group paused and listened.

"Maybe to keep the birds away," Lynwood concluded. "Yeah. A radar for the birds."

"Sometimes we hear a sound like marbles at night, too. A whole bunch of marbles being turned over. Is that when they dump the coal cars?" the Asian guy asked.

"Now listen, we only dump those during the day. Not at night. It's gotta be somethin' else," Lynwood started.

"I'm not complaining, I'm just saying I hear it."

"Yeah, we're not complaining," the yuppies jumped in a little defensively.

We took the elevator down from the roof to the bottom floor of the plant. Here the ash from the generators was gathered, lumping into an almost iridescent gray rock of ash. Lynwood nudged a lump with his foot. "Some cities use this stuff as road fill."

We stepped out of the plant into the yards and walked up to a hut made of steel siding, watched over by a kid in his midtwenties who looked very different from the type of employee we had encountered in the plant: grubby jeans, a T-shirt, and a battered ball cap. Mirant had partnered with a local wildlife organization to introduce sturgeon into the Potomac. Along the perimeter huge tubs contained these fish at different stages of maturity. The unfiltered water they swam in was drawn right out of the river, and so far, the sturgeon were alive, slipping around in their confinement. I imagined their bellies sliced open for black caviar to sell to restaurants and fish markets on the waterfront. I wasn't sure I was willing to eat anything from the womb of a fish that filtered Potomac River water in and out. Any pollution the plant had added to the river over the years really was *just a drop in the bucket*; certainly many other companies had done their part.

Sturgeon wasn't the only wildlife-friendly initiative on board for Mirant. "And we're working on a project to develop a nesting area for bald eagles," Mike Stumpf had been careful to add at the end of his presentation.

22

Yes to Electric Reliability

Mr. Ralph M. Hunt writes: "I support the call for reasonable solutions to provide reliable, affordable energy while also protecting our air quality, so I finally decided to get involved."

Alice Bertele writes: "This letter is to share my fear and frustration about how our City Council is gambling with our environment, health, and more than $3 million just for the sake of political rhetoric ... Instead of grasping reality, the City Council is playing politics by fighting the Potomac River Generating Station and in doing so, they are actually preventing innovative improvements to our air quality that the plant owner, Mirant, has proposed."

The letterhead on both notes bears the imprint Bright Ideas Alexandria, a firm hired to promote a positive image of Mirant in the community and garner support for their proposed stack merger. Ralph, for all I know, may be fictional—no Ralph M. Hunt is listed in the Alexandria white pages. A seventy-year-old resident, Alice Bertele exists, but I wonder how well-informed she is about the lawsuit between Mirant and the city of Alexandria. A brochure sent out to local residents depicted a father swinging his daughter in front of a backdrop of blue sky. "A Plan for Better Air Quality" was printed along the bottom.

I accepted the invitation in one of these letters to attend a public hearing before Virginia's Air Pollution Control Board. The Potomac River Generating Station's operating permit was due to expire a month later in June 2007, and both plant employees and Alexandria residents would have the opportunity to voice their positions. Mirant was sponsoring an open house in a nearby Tex-Mex restaurant, where plant employees mingled with local residents to talk about the plant and dig into plates of flautas, chicken wings, and quesadillas. Employees bunched together in the leather booths at the restaurant. Most were men, lounging in ball caps and golf shirts. I was one of the few from the community. Another city resident had nothing better to do. One was unemployed. One was curious. The majority had no clear sense of the issue Mirant was recruiting us to defend.

Within an easy reach of the flautas, I sat across from Misty Allen, the young lawyer who had presented with Mike Stumpf at the plant's open house. Just out of law school, she was friendly and disarming. She smiled often and nodded as if she were confiding in a few close friends. "If people build houses around the airport, people complain about the sound when the planes come over. It's about the right to do business. Mirant was here first, and the community was built up around. It's very American to claim that something is somebody else's fault."

"I'd just rather they burn the coal here than do it in West Virginia. You lose energy over transmission lines," responded the man sitting next to me. He was good looking and about my age. New to the area, he was the one still looking for a job.

He had a point. The most recent statistic, compiled in 1995, estimated a total loss of 7.2 percent for electricity distribution and transmission in the United States, which seems minimal until you realize at the rate of supplying half the needs of the energy grid, this equals 37.5 million tons of coal. I too would rather we suffer the consequences of our own energy consumption. As the Potomac River plant served Maryland and D.C. residents and the federal government, Alexandria was paying the price for somebody else's energy. Dominion Resources, a company headquartered in Richmond, Virginia, provides Northern Virginia's power supply, and somewhere another Virginian community is breathing the byproduct of our electricity.

Mirant is fond of saying it's been around since before the city was built up around it, and this is partially true. Although the plant was constructed in 1949, the current owner obtained the facility in 2000. When Mirant bought the generating station, it was well aware that it was grandfathered. The Clean Air Act was passed in 1970, placing caps on the noxious emissions of sulfur dioxide, nitrogen oxide, particulate matter, ozone, carbon monoxide, and lead. At the time, coal-burning plants contended that requiring preexisting facilities to comply presented an unreasonable financial burden. Legislators, believing that eventually these plants would be retired, granted the exception.

Instead of abandoning the old facilities, the coal industry has continued to operate them. Building a power plant is an expensive endeavor; it takes years to recoup the initial investment. Because the Clean Air Act exempts grandfathered plants from pollution controls, these older plants are also less expensive to run. Consequently, for the equivalent of a matter of pennies per household per month, half of the coal burning plants in America are still spewing the same amount of pollution into the atmosphere as they did prior to the environmental movement. Coal fired plants are responsible for emitting 22 percent of the nitrogen oxide, 60 percent of the sulfur dioxide, 33 percent of the mercury, and 40 percent of the carbon dioxide in our air. A whole century of progress is yet to be implemented in the electricity industry, and half of all Americans breathe air that doesn't meet ambient air quality standards.

Mirant knew very well that buying the Potomac River Generating Station was a cost-saving endeavor. It was a business decision based on financial considerations, regardless of the unhealthy consequences for the surrounding residents. Compared with new coal burning plants, a plant built in 1949 is ten times more polluting. Since the Clean Air Act was implemented, sulfur dioxide emissions have decreased by 77 percent; nitrogen oxide, by 60 percent; and particulate matter, by 96 percent. As a result, the EPA estimates the US public has saved forty-two dollars in health care costs for every dollar spent on pollution controls for every year from 1970 to 1990. This is a cause for celebration and a case for effective legislation; however, it also reveals the sharp correlation between health and air quality, and how much more remains to be done. It demonstrates the money required to retire old facilities are dollars well spent. It's an investment in public health that proves to be ultimately less costly to consumers who bear price of upgrading anyway.

I walked over to the hotel where the hearing was being held, conversing with the water engineer for the plant. A tall polite, well-mannered man, he explained his role at the plant: "I do all sorts of chemical tests and adjust the pH before the water goes back in the river."

"Are any chemicals released into the river?"

He looked at me suddenly and suspiciously. "Well, nobody's perfect."

At the hearing, I situated myself behind the handsome man from the restaurant, who had taken a seat next to an elderly couple.

"Is this meeting for or against the plant?" the woman asked him, her voice cracking a little.

"It's a hearing, I think, for both sides," he responded.

"We've just come to see what the meeting's all about. Do you come to these often?"

The parties for and against lined up on either side of the hotel meeting room. Mirant's supporters wore "Yes to electric reliability" stickers and yellow cardstock light bulbs on golf shirts. The residents of Alexandria wore suits. From the looks of it, it wasn't just an argument between public health and business; it was a conflict divided along class lines. Alexandria has become more affluent over the years. Old Town, just down the river from Mirant, is one of the most expensive and picturesque neighborhoods in Northern Virginia. Retired people and yuppies who work in D.C. stroll down the cobbled sidewalks with their dogs, past the best dining in Northern Virginia. Along the waterfront on summer days, people eat ice cream and stretch out on blankets on the grass. Alexandria supports one of the few artists' communities in the greater Washington area and puts on a yearly art festival. During the winter, Christmas lights on trees brighten King Street, the main commercial drag.

The majority of Mirant's employees had worked at the plant for at least seventeen years, and many had lived in Alexandria's neighborhoods. By the time of the hearing, a number had moved out. Usually this sort of trend in an industrial area indicates that the financial prospects of industrial employees have improved, but that is not the case here. The majority of Mirant's workers are blue collar, having entered the energy industry right out of high school. Many of the new residents of Alexandria are white collar and rent or own homes whose property values have skyrocketed out of the pay range of the employees of the neighboring plant.

Among plant workers and city residents sat young adults dressed in grunge: interns and new hires from local environmental groups. Here and there were men in the most tailored and blackest suits, lawyers from the District representing the national government.

Vice-Mayor Adella Pepper was one of the first to speak. A member of the Mirant Community Monitoring Group, she had been heavily involved in the issue since Mirant was shut down in 2005 in response to a warning from Virginia's Department of Environmental Quality (DEQ) when it was discovered the plant was emitting fifteen times the amount of legal SO_2.

Del, a small woman with pale orange hair, seemed like the sort of person who led a pep squad fifty years previously at one of the local high schools. Emanating spunk and intelligence, she started into her speech with calculated vehemence: "I've been a member of the Alexandria City Council for the past twenty-two years, and in all my years in public service, I have not seen a greater public issue than the one that is before us today—that is, ensuring that the operations of the Potomac River power plant are no longer injurious to the health of Alexandria's residents and guests. There are about twelve thousand people living within a mile of the plant and twenty-five thousand people living within one and a half miles. I've never seen such an aversion on the part of a corporate citizen to do what is right for its neighbors and community."

For a long time the only evidence of the plant's impact on the nearby residents was anecdotal—mostly complaints of stinging eyes, asthma, and cancer. Like the residents of Sundial and Sylvester in West Virginia, the citizens of Alexandria were tired of sweeping black dust off their porches and windowsills. In 2001, in response to the concerns of community, the city starting monitoring the plant.

Del continued, "A test from the plant itself revealed, aside from the chemicals and pollutants released by the smokestacks, that twenty-nine tons of fly ash and ten tons of coal dust were escaping from the plant each year into the surrounding air and community."

After Mirant's test results were released, the city hired Dr. Jonathan Levy from the Harvard School of Public Health to conduct a study of the health impacts of the plant. Levy's findings are notable. Particulate matter, which only accounts for a portion of the plant's emissions, was causing 59 premature deaths, 66 hospitalizations, 870 emergency room visits, and 4,600 asthma attacks per year in the greater Washington, D.C. region. During the year in which Levy released his findings, Mirant exceeded its permit limitations for NO_x (nitrogen oxides that are created during combustion). Alone, these chemicals are not hazardous, but they combine with water in the air to create nitric acid, which results in acid rain and ozone in the lower atmosphere. Ozone is necessary in the upper atmosphere to shield the earth from the sun's most intense radiation, but a high concentration in the air we breathe can destroy lung tissue, causing emphysema and bronchitis and aggravating heart disease. According to Del, after these initial findings in 2005, studies by the plant and the city indicated violations of National Ambient Air Quality Standards and Virginia Toxics Standards.

Of particular concern to Alexandria residents is downwash from the Potomac River facility to Marina Towers, a high-rise built nearly on level with the smokestacks. During certain weather patterns, the emissions from the stacks are blown into the side of the apartments instead of being dispersed into the air above. Consequently, the particulate matter, sulfur, and other toxic chemicals released by the station are inhaled by the residents of the Towers and the community under its shadow in greater concentrations than the output levels of the plant would usually indicate. Before the meeting, Misty Allen had asserted that Congressman Jim Moran (D-Virginia, and a former mayor of Alexandria), who had been vocal in his opposition to the plant, had a vendetta against the company only because one of his lady friends lived in the Marina Towers—a view of the plant was the backdrop for their rendezvous.

Del continued, her energetic delivery undiminished by the length of her speech: "It has been the persistence of the city, its residents, and its staff to bring all of us here today to push for strict operating controls, controls which should have been in place years ago. This should not have been left for the city to do … The burden should always be on Mirant. It is a travesty instead that Mirant reminds us that the plant predates many of the residents in the neighborhood if this in any way justifies subjecting anyone to health risks. Mirant also justifies its approach by accusing the city of being cajoled by a vocal minority of residents."

The next to speak was one of these vocal residents, the kind that both Alice Bertele and Ralph H. Hunt would deride. A man in his thirties or forties, David England made little attempt to veil the anger in his voice: "I speak on behalf of thousands of friends and neighbors who are tired of their eyes burning, their lungs feeling tight on what ought to be beautiful days jogging on the Mount Vernon River Trail, of spending time in beautiful, historic parts of Alexandria. We are tired of particulate matter having to be swept off of windowsills and off of our property. The Department of Environmental Quality has failed us on this. Their responsibility is to protect the environmental quality of our community, not to try to find every way possible for a plant to operate. Thus, we have turned to you, the Air Board. The message that I want to leave with you is that this is really a balancing act between public health and money. There are a number of steps that can be taken to protect the health and safety of Alexandria, but they cost money. I have here Mirant Corporation's 2006 report: $918 million dollars. When they say that they can't afford to support the city's proposal, when they say they don't have the money to implement the kind of measures that will protect the health and safety of this city, it just ain't true."

His argument was answered by Bob Driscoll, the CEO of Mirant Mid-Atlantic Operations. Like the lawyers from D.C., he wore a black suit and spoke in an expressionless voice that evoked authority: "This is an issue of balancing the social value of reliable energy and Mirant's right to operate the plant. Mirant has demonstrated that operations and the investments we have made have complied with and supported the protection of human health in the greater Washington, D.C. area. The monitors have shown that the plant operates consistently well below any max limitations. Furthermore, this is not an issue of money. We have spent so far tens of millions of dollars on and are prepared to spend thirty million dollars more on our stack merger project."

The appeal to electric reliability is not new; it's consistent with the rhetoric of coal plants and the governmental agencies that have supported them since ambient air became a public health concern. In 1970, when the Clean Air Act was implemented, power plants complained that both these restrictions and the EPA threatened "the safe, reliable, and efficient operation of energy production across the country." Given the health hazards of coal burning plants, "safe" is probably a misnomer. Nor was "efficient" quite the stretch then that it is today. Currently, grandfathered coal plants are an ineffective way to burn coal; newer coal plants are 10 to 20 percent more efficient. Reliability is a coal plant's claim to fame: electricity delivered steadily by antiquated giants.

Driscoll continued with Mirant's most convincing argument: the Potomac River Generating Station is necessary for electrical reliability and redundancy in downtown Washington, D.C. After the Virginia DEQ shut down the plant for emissions violations in 2005, the Department of Energy issued a rare emergency order to resume operations. Three of the five broilers revved up, spilling toxins into the nearby neighborhood. The DEQ order was about to expire in June, a month after the hearing. In order to ensure the future energy needs of the capital, Pepco was installing two transmission lines under the river from two of its other coal burning plants, a project which was slated for completion at the end of July 2008. This was why the argument between the City of Alexandria and Mirant involved a third player, the federal government.

Mirant proposed two solutions to mitigate its pollution problem. The first was the use of trona (trisodium hydrogendicarbonate dihydrate), a mined mineral used to make baking soda that helps power plants collect particulate matter and unwanted gases. Mirant asserted that this would reduce its output of pollutants by 45 to 50 percent. To mediate the issue of downwash with Marina Towers, Mirant proposed the stack merger project heavily advertised by Bright Ideas Alexandria. Combining the output of the five stacks into two would theoretically create sufficient air pressure to force the emissions higher into sky, creating more "virtual" stack height.

Next to testify was Lara Greene, a toxicologist and chemist from the Massachusetts Institute of Technology (MIT). She was losing her voice and waved away offers of water, claiming it wouldn't help. She began by stating that Mirant had given her and her colleagues a grant to study the problem. As a scientist employed by a power plant, she chose her words carefully in her testimony: "Anybody who lives in Marina Towers knows the stacks are far below the height they should be. When Marina Towers was built the idea of stack height didn't exist yet. Since the 1980s, engineers have known that unless a stack is at a certain height, it is capable of creating a downwash situation for nearby residents in high towers. I would urge you all haste to increase the stack height, virtually in the stack merger, and I would also urge an actual stack height increase."

This statement contained an admission that perhaps that stack merger would not solve all the downwash problems. A year prior to the hearing, Mirant had applied for and been granted permission by the Federal Aviation Administration to raise the stacks by fifty feet. Due to the expense of this project, Mirant decided to propose a stack merger instead.

Greene continued with another point of contention Alexandria had with Mirant, SO_2 emissions: "It is the case that the primary National Ambient Air Quality Standards for sulfur dioxide are not sufficient or protective under a very special set of circumstances which may or may not exist here. Although people without asthma—and many with asthma—can tolerate a reasonably high level of sulfur dioxide, there's a subset of asthmatics who are very sensitive to SO_2. Now, this only happens upon hyperventilation, such as when somebody is exercising, but if you happen to be a SO_2-sensitive asthmatic who is exercising in the presence of SO_2, within a few minutes, bronchial constriction takes place."

When Greene mentioned this subset of asthmatics, I was reminded of my personal stake in the lawsuit. I live seven miles from the plant by car (and four as the crow flies) and belong to the region evaluated in Levy's study. Asthma is a problem that I have dealt with since I was fourteen. Long-legged and tomboyish, I kept up with the fastest boys in grade and middle school. Ever since, running has been a challenge, and exercise is something I have to work into slowly. Maintaining my aerobic health is essential, a challenge when allergies increase my propensity for illness and aggravate my already serious case of asthma. Even in my early twenties, when I worked for the Forest Service and was strong enough to carry a ten foot fence post on my shoulder up a hill, I had to take breaks in long hikes to keep from passing out.

One of the pioneers in studying the health impact of power-plant emissions, C. Arden Pope, worked as an economics professor at Brigham Young University when my Grandpa Thomas was still one of its vice presidents. Between 1985 and 1988, the Geneva Steel plant—located near where I grew up in Orem, Utah—temporarily closed over a labor dispute. Pope took the opportunity to evaluate the claims of many mothers in the area that the plant was making their children sick. He studied hospital admissions during this period and found that admissions for pneumonia, pleurisy, bronchitis, and asthma were between 50 and 86 percent higher for children while the plant was operating, and during the worst weather conditions, they were up to 47 percent higher for adults. It has been documented since that children who live near power plants develop smaller and weaker lungs.

I don't know if I belong to the group of SO_2-sensitive asthmatics or if my asthma is the result of my proximity to the Geneva Steel plant in my childhood, but I do know what it feels like to hyperventilate for hours and wake up the next day aching from the byproducts of anaerobic respiration in my muscles. I know what it's like to be carried home from a basketball game because I can't walk or to lose my vision temporarily because there is not enough oxygen in my blood to operate my brain. It's a big deal for me, one that my doctor advised me could be deadly.

Next, Greene addressed Mirant's plan to use trona to control particulate matter, which was first proposed in 2005 after the plant temporarily ceased operations. The plant's original experiments with trona were kept secret from the public, which created suspicion among residents. Using trona was a new technique, and the plant had considered submitting their approach for patent. Earlier at the open house, Misty Allen had explained that trona was "just like baking soda" and "perfectly harmless." This argument probably holds up if you were proposing to scrub your toilet with trona, but the repeated inhalation of baking soda could be another matter.

Greene admitted, "Trona is not the best way to control SO_2, but it is a feasible way to control SO_2, and it seems like it will be feasible here. Now trona has a downside; a miner or miller who is working with trona can experience irritation ... In Alexandria, the amount of trona that would escape is such that at most the impact of half of a microgram."

A report released by the EPA in 1977 under the Carter administration indicated that scrubbers were the best way to eliminate SO_2 from power plant emissions. EPA Deputy Administrator Barbara Blum claimed: "The only process with high sulfur-dioxide removal efficiency widely available now is scrubbers." As of 2004, only two-thirds of America's 420 plants had scrubbers installed, even though they have been shown to reduce the SO_2 emissions by 95 percent.

The last part of Greene's testimony was the most passionately delivered, despite her raspy voice, indicating that there is an aspect of a scientist that no company can buy: "Of all the pollutants that we have to deal with in the East Coast, the one that matters the most is ozone. I am horrified that cars do not have mandatory mileage controls. I am disgusted that ozone has not become better controlled in our society, especially on the East Coast, and I'm frankly mystified that nobody seems to be focusing on ozone. Ozone is a problem for all of us whether we have asthma or not, whether we're young or old. So when people have burning eyes and irritation from the short-term effect of ozone, and probably other oxidizing pollutants, I would urge you also to think about the oxidizing pollutants, not just focus on reducing SO_2." Greene didn't connect this speech to the plant directly, but implicit in her testimony was the fact that larger pollution issues were at stake.

The D.C. contingent was up to bat next. Kevin Kolevar, director of the Office of Electricity Delivery and Reliability in the DOE, spoke first: "We are interested solely in having reliable energy to central D.C. protected. That is why the DEP [Department of Environmental Protection] issued an emergency order, one rarely used, in December 2005 that requires the operation of the Mirant power plant under certain limited conditions."

The federal government sought a permit from the Air Pollution Control Board only until the completion of the two Pepco lines; they promised no further intervention afterward. The City of Alexandria was mistrustful, speculating that the federal government would reverse this promise once the energy demands of the District increased. For the previous two years, in order to comply with the Northern American Reliability Standards for electricity, the government had violated the Clean Air Act.

From Mirant's perspective, the involvement of the federal government had bought them time. At this particular hearing, it was their trump card to ensure the continuance of the plant after the Pepco lines were completed. Because Bob Driscoll and Kevin Kolevar had established the necessity of the plant, the plant employees took a more humble stance: an emotional appeal to the American values of the well-meaning workman. Mike Stumpf led off with a slightly different air than in his presentation at Mirant. His voice took on more of a Virginia regional twang: "I am one of the 122 people who work at the Potomac River plant: finely skilled electricians, chemists, instrument repairmen, mechanics, and operators that work in this area and want a clean environment, just like everybody else."

At this point a number of Alexandria residents yelled, "How many of ya live here?"

They were countered by a number of Mirant's employees on the opposite side of the room.

Mike continued: "We take our responsibility seriously because we know we can make a difference. I've worked at the plant since 1998 and can say with confidence that our environmental performance under various ownerships has improved in every area, with dramatic reductions in NO_x and SO_2. The site is cleaner than it ever has been before ... I hope you recognize that achieving this level of performance would not be possible without a workforce dedicated to continuous improvement. We are proud of what we do and have reached out to the community more than ever before ... Our goal remains to operate in a way that protects public health and the environment while simultaneously providing electric reliability to the region."

In fairness to Mike Stumpf, operating a plant built in 1949 according to any level of current pollution standards would be a formidable undertaking. The relevant question would probably not be whether the workers at Mirant are doing their best with what they have, but whether a coal burning plant built in 1949 should still be generating electricity.

A number of plant employees followed. Both Victoria Gross, a large African American woman, and her husband worked at the plant. She lived in Maryland, but previously her family had lived close to the plant for many years. All of them, she testified, were healthy: "What most concerns me is that our adversaries don't care about existing together as neighbors. They don't seem to care about working together to find solutions. To problems that would benefit all parties concerned, someone has to be the voice of reason. This should not be a dictatorship call. As surely as I stand before you today, a healthy, vibrant individual, I urge you to adopt a balanced operating regimen for this facility."

Lynwood Reed, my gentle dispositioned guide at the plant, spoke next. During his testimony, I realized the reason for his defensive reaction to my group's commentary on the beeps. "We not only have problems with the emissions of the stacks, but we get a lot of complaints from the city about noise and particulate matter, and on every circumstance, we have responded, and we have mitigated those problems, and we will continue when problems come up to address them, and we have been addressing them. We are a company that's truly cares about our neighbors. After fifty-eight years as a servant of the Washington metropolitan area, we have offered reliable electric power using American's largest fossil fuel source, coal."

Dexter Hanford was the most animated and the most accusatory in his language, outdoing his colleagues in the image that Mirant was trying so hard to project: "I'm one of them employees from Mirant company. During the past couple years, I been the target of a lot of false accusations. Activists and other forces that wish to close the plant have labeled me a polluter—as one that makes the children sick. Obviously they don't know me. I have worked in operations for Pepco Mirant for eighteen years, and I am the father of two children, and I'm a h-a-a-rd worker."

"My job is to help provide reliable electricity to the power grid so that any of you can enjoy his or hers air conditioning during the summer or have heat in the winter. This is a job I take great pride in. I do this while operating with an overload of environmental guidelines consistently. One thing that I can't do is look anybody in this room—or anybody else for that matter—in the eye and say that not once have I been told by this company to violate any environmental guidelines. I also want to say, as an employee of eighteen years, I have had the opportunity to see the retirement of dozens of plant operators, all of which retired in excellent health after serving for more than thirty years. No one can say they've seen employees walking by with respirators; that is because we are very comfortable breathing the air in our environment. Act reasonable and allow me to continue to work to support my family."

Geneva Steel was one of the leading employers of the community I grew up in. I remember the fear that circulated in my neighborhood and church congregation when the company began to decline and lay off workers. Geneva Steel finally closed its operations in 2001, and the loss of this industry had an economic impact on the area for several years. Currently, this no longer affects the prosperity of the community, but Utah County's air quality still ranks among the worst in the nation. The steep mountain ranges that rise on all sides of Utah Valley trap pollution during summer and winter inversions, and on days when the air quality reaches red levels, all are advised to stay indoors. Mirant is quick to assert its right to operate, and although it owns the land that the plant occupies, Mirant does not own the Potomac River or the air. Water and air feed into global cycles and impact our collective health.

In May 2007, Virginia's Air Pollution Control Board ruled in favor of the city and issued a temporary operating permit based on the tighter restrictions proposed by Alexandria. Mirant sued and lobbied the state government to remove the power of the Air Pollution Control Board and the Water Control Board to issue permits. The Potomac River plant operated while Pepco finished its lines and continued to operate, no longer necessary to the electrical reliability and redundancy of the D.C. area, but in keeping with its right to do business. Alexandria persisted in its suit, and in 2008, the city reached an agreement with Mirant, locking in more stringent emissions standards and requiring the installation of baghouse filters to reduce particulate matter. In order to make these changes at the plant, Mirant pledged $34 million. This was a temporary victory for the city, but it hasn't lost sight of its ultimate objective. This is where the workers' testimonies had Alexandria pegged. Local residents will not be satisfied until the Potomac River Generating Station closes operations permanently.

In a situation where there are so many disparate concerns, perhaps the only democratic solution is the numbers game. We could tally up the number of exercising asthmatics in Alexandria and compare that with the number of employees at Mirant. We could count the number of residents in Alexandria intent on closing the plant and compare that with the number of people in Maryland who use the electricity Mirant produces. We could also compare this amount with the number of Appalachian folk employed, displaced, or poisoned by the mine from which the coal used by Mirant is dug, and then we would have to combine this figure with the number of Appalachians impacted by the coal processing plant.

There are some who still assert that Mirant may be necessary for the electrical reliability and redundancy of D.C. as demands for electricity increase. We could compare the effects of D.C. going black for a day—the local effects on the hospitals and the local and global effects on the economy—but then, of course, we would have to factor this against the worldwide impacts of the CO_2 released into the atmosphere.

In the public hearing on the Potomac River Generating Station, there wasn't much talk of alternative energy. It wasn't the immediate issue at hand, but for some residents of Alexandria the coal burning dilemma may be as simple as burning it somewhere else. "Somewhere else" would inevitably fall on our less prosperous neighbors, who already suffer the human costs of digging and cleaning coal, and their rates of lung cancer would be elevated accordingly. "Burning it somewhere else" is a solution that's merely hiding the problem, ignoring the coal in our veins: a refusal to pay the actual price of our lifestyle.

Mirant can be a little defensive. But in a way, one can hardly blame them. Energy can't be stored except in batteries, and it is generated on an at-need basis. The truth of the grid is that demand directly impacts supply. There is no middleman between our energy use and the power plant;

the grid is only a highway. The moment we turn on a light, we tap into power from the grid. More energy is generated to keep the grid at equilibrium, more coal is burned, more coal is dug. In the last ten years, energy demands have gone up by 40 percent. The average American uses twenty pounds of coal a day. *Down there where the coal is dug it is sort of world apart which one can quite easily go through life without ever hearing about ... Practically everything we do, from eating an ice to crossing the Atlantic, and from baking a loaf to writing a novel, involves the use of coal, directly or indirectly ... We are capable of forgetting it as we forget the blood in our veins.*

Or in the words of contemporary journalist S. C. Gwynne: "When you turn on your television set, air conditioner, or dishwasher, you are doing something that is at once mechanical and moral. In order to power your appliances, you must summon electricity from a vast sea of energy ... Generating plants across the [nation] powered variously by coal, gas, nuclear fission, wind, and water dump electrical current into this grid, and by flipping a switch you engage them all. No matter what your politics are, how much of an environmentalist or conservationist you might be, or whether you actually give a damn where your electricity comes from, you're complicit."

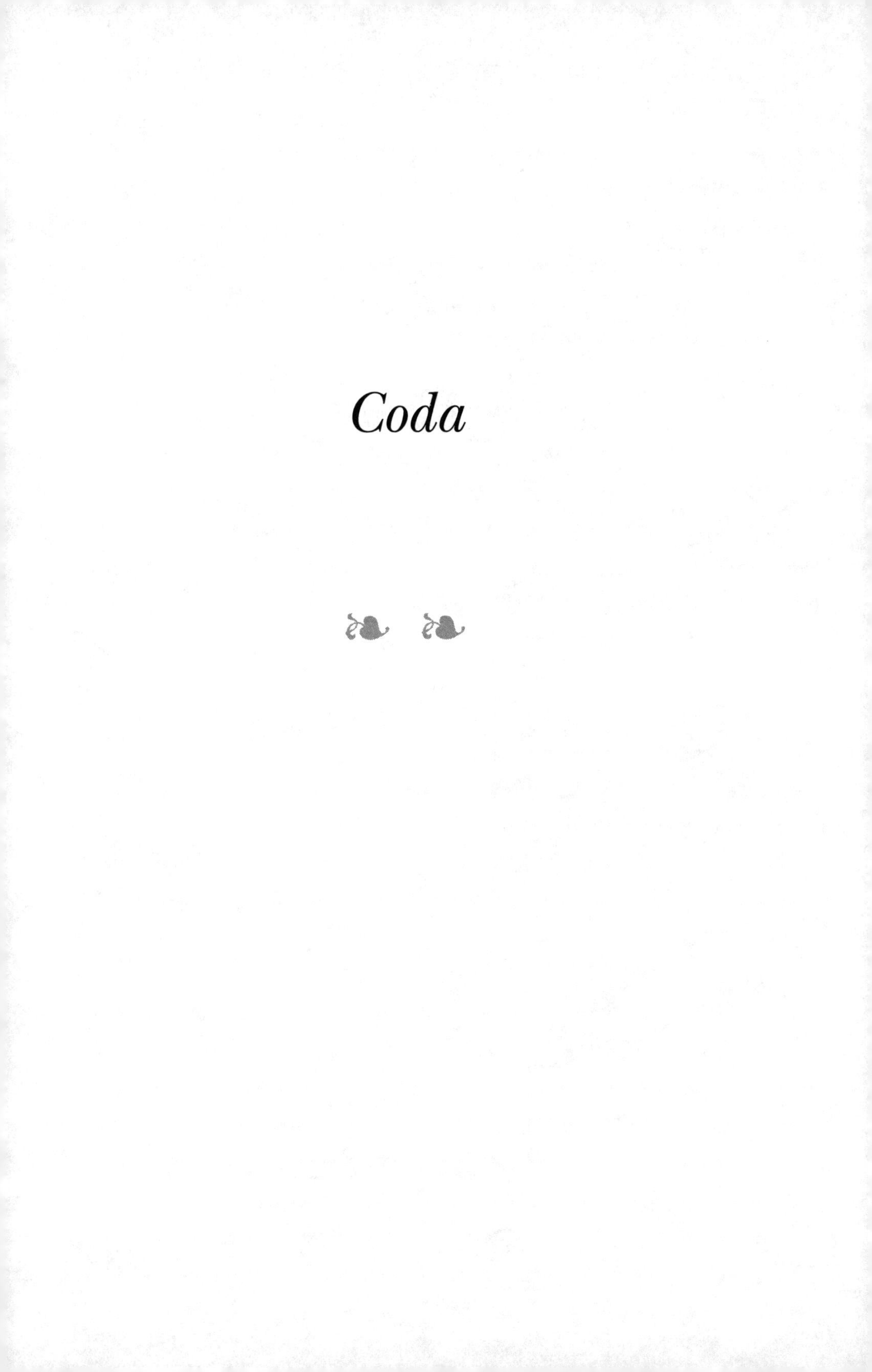

Coda

23

In the Bowels of the Earth

The Wheat Cap Lamp looks very similar to the cap lamp designed in 1912 by Thomas Alva Edison. A wide-faced light clips onto a hardhat, and the wire is fed over the hat to where the battery is held on a belt. The lamp can provide twelve hours of light for each battery charge, and the battery is rechargeable five hundred times. Each bulb has two filaments that add up to 550 hours of bulb life. This is not the type of battery-powered lamp that waited in the storeroom in March 1924 when my great-grandfather Zeph sent in his crews to Castle Gate Mine No. 2. The batteries of those lamps were heavier and worn as a backpack. Zeph would have called them *Edisons*. In 1924, any electrically powered lamp was cutting-edge technology, employing two late nineteenth-century developments: the rechargeable battery and the light bulb. Why the Utah Fuel Company didn't issue the lamps immediately was a question Zeph often agonized over after the accident because this invention became a life-saver. The Castle Gate explosion was not the only disaster battery-operated lamps could have prevented. At the time, open-flame lamps caused 53 percent of mining explosions.

Not long after Edison developed his cap lamp, inventor Grant Wheat developed a cap lamp for the Koehler mining company in 1919. By the time Edison received his patent in 1928, the US Bureau of Mines had already approved Wheat's lamp. In 1935, a deal was struck with a British manufacturing company, and this reliable miner's light came into frequent use in the 1940s and 50s. Currently, the Wheat Cap Lamp is used by 85 percent of US and Canadian miners. You could call it karma. Edison's ample means allowed him to purchase the first patent for the light bulb

Erin Thomas, 2008

Skyline Mine

developed by Henry Woodward and Matthew Evans in 1875, and for Edison's improvement on the Woodward-Evans invention, history has dubbed him the Father of Electricity. Due to Wheat's improvement on Edison's invention, when I pulled on my gear to enter the Skyline Mine in 2008, I attached a Wheat lamp to my cap, and not an Edison.

The black rubber boots from the guest locker pinched a little, suctioning to my calf. I stuffed the legs of the blue disposable coverall into the top of the boots, because the floor of the mine was likely to be wet. A heavy leather belt hung on my hips, with a pouch for the battery to my Wheat Cap Lamp. When I visited her in Scofield, Ann Carter was kind enough to invite me back to visit the mine where she works in accounts payable. The Skyline mine operates just up the canyon from Winter Quarters, and it began operations in 1981.

A safety trainer equipped me with two self-rescuers from the checkout room. Since the Sago accident, MSHA has required each miner to take two into the mines and leave one within a fifteen-minute walking distance. Each one provides an hour's worth of oxygen. The trainer showed me how to inspect them for functionality, and I realized how much it is a miner's responsibility to ensure his own safety. In the Sago accident, only three of the ten self-rescuers were functioning. This was avoidable.

One enters the Skyline Mine through a level. There is no shaft, just a hallway that angles down to the heart of the action, hundreds of feet underground. The Skyline intake tunnel was not the arched burrow I used to envision. It was more geometrical, a rectangular hallway, causing me to recall the Chinese miners run out of Carbon County. This was the sort of tunnel they would have dug. My guide for the day, Stan Christianson, drove us through it in his truck. He was tall, broad-shouldered, and mustached, a gentleman-miner practiced in promoting his trade to those from the outside.

The roof and walls were not the slick black of my trip to Big Pit, but a chalky brown color that reflected the light murkily. This was the result of rock dusting—discovered to be a fire deterrent after the Castle Gate explosion. Perhaps some good has come out of the accident for which my great-grandfather Zeph felt so responsible. As a result, fewer miners have died since. Stan told me the air in the mine is required to be from 60 to 80 percent nonflammable.

A rope line hung from the roof, flagged at increments. Stan pointed at it, as he would point to a variety of things throughout the time I shadowed him, explaining yet another possible violation of the Mining Safety and Health Administration's standards. This was also why the floor was so level. The rope was "one of the new safety guidelines. You can get fined if it droops too much. This year so far, MSHA has levied $400,000 in fines. Much more than any year before," Stan explained. Each mine is required to submit to forty days of inspection per year; "some inspectors were friendly, but others could be really cranky."

After Sago, Congress passed the Mine Improvement and New Emergency Response (MINER) Act of 2006, in order to increase safety guidelines. According to Stan, from the operators' and the miners' perspectives, these additional safeguards made a mine nearly impossible to operate. One self-rescuer added weight to an already heavy outfit—rubber boots, the Wheat Cap Lamp, and battery—two seemed like overkill. When I visited Skyline, self-rescuers were stashed throughout the tunnel along with two contraptions the miners dubbed "EBOs," or easy bake ovens. These *refuge chambers*—long metal boxes that resemble trailers—were equipped to provide oxygen for eleven men over a period of thirty-six hours in case of an explosion. Stan informed me that each cost $125,000. "They don't ask," he said referring to MSHA. "They just said, 'Put them in.' But I tell you, if there was an explosion, and there was any chance of gettin' out, they'd all chance that instead of going inside that thing."

From a legislative point of view, these new technologies "represent the most significant improvement in mine safety technology and mine safety practices in three decades." Despite the financial burden of the new regulations, Skyline seemed to be doing well. Most miners made $65,000 a year, working three thirteen hour shifts that were always separated by a couple of days off in between. Some who worked overtime could clear as much as $100,000. Stan was near retirement and felt confident about his benefits; he had money tucked away in a 401(k) retirement plan and a full-service medical plan he could pay to have extended beyond his retirement.

Another part of the MINER Act involves new safety procedures to track miners throughout the mine. Each mine is divided into zones; when crossing into a new zone, the miners radio an operator, who maps out their new location. Every miner has a number. I was visitor number 6, a designation I wore as a circular metal tag attached to my self-rescuer. This tag reminded me of the unidentified miners killed in the Winter Quarters mine, now reported to haunt its premises. Although logically I knew there was little chance of danger, my heart had been beating fast all morning, and I was short of breath. I reflected grimly that they would have no trouble identifying my body.

Stan stopped his truck at the end of the intake tunnel, and we stepped into the paused operations of a longwall. A longwall is a remarkable piece of machinery. Skyline's was nine hundred feet long. Hydraulic jacks support the roof, creeping forward as the shearer spins, scraping coal off a thirteen foot tall face. The jacks are set side by side, and can weigh up to forty tons each. Stan walked across the foot jacks to talk to the longwall operators, and I followed him.

"Don't step in front of the jack," he warned me. "It could move forward and take off your foot." Longwall mining replaced the room and pillar method in the 1950s. After the jacks shudder forward, the wall behind is left to collapse, causing the land above to sink. This can result in changes in the water table and other instabilities, but the longwall makes it possible to retrieve 80 percent of the coal in a mine in contrast with 60 percent using older methods. It is completely automated, extracting ten thousand to thirty thousand tons on a typical day. A scraper sends the coal down a conveyor belt, which moves the coal through a crusher to standardize the lump size, and then sends it out to the tipple, where it lands on the top of a pile. There are no shovels, picks, or trams. No boys sort the coal from the rubbish or the lump from the bedlam.

One of the miners changed a bearing in one of the jacks, and then the roar of the machine sent the scraper spinning up and down the length of the wall. Fewer men work a mine than ever before. At Skyline, there are no more than twenty-five to thirty men in the mine at a time. Only two men operated the longwall, standing on the footpads of the hydraulic jacks under the protection of the arms that held the roof above their heads.

As we left the longwall, Stan pointed to the belt: "We get fined if we get any accumulation under the belt. Coal can spark from the belt if it gets too high. But accumulation," he gestured at the few bits of stray coal several feet under the suspended belt, "that's ridiculous."

Stan had his lines down pat. From what he told me of his appearances at public educational forums, he did quite a bit of promotion for the mine and for coal itself during his free time.

"These environmentalists, they care about the environment, I understand that. But we've got ten disturbed acres here. We saved topsoil and boulders for cosmetic reasons, and when we leave, you won't be able to tell."

I nod. I knew that the bonds the companies posted for the reclamation of coal mines in Utah were taken seriously. And Stan was right about the impacts on the land. There is no thick forest on the lowlands of the Rockies, and it takes few years for the brush to grow back. There were no displaced towns or filled-in streams. Skyline could be considered a fairly benign operator. But it is owned by Arch Coal, the second largest coal mining company in the United States—the same Arch that refuses to save topsoil in West Virginia to spread over MTR sites because it is too expensive.

Stan parked, and we entered a part of the mine that was darker. Although the ventilation system ran along the roof, I could feel the air I breathed inside my lungs. It pained them slightly, similar to the ache of inhaling cigarette smoke. We walked into a tarped off section furnished with a dinner table, and Stan laid out some the plans to show the foreman. They were digging a new face, a process that required other sorts of heavy machinery.

The ground was wet, and my rubber boots splashed in the puddles. I suddenly felt awkward tromping after Stan with all the heft of my two self-rescuers and the Wheat Cap Lamp, which was shining out in front of me. Mines have rules governing cap lamp etiquette, designed to avoid blinding coworkers and causing accidents. Twisting a knob on the side of the lamp narrows the light to a powerful beam, which a miner should never aim directly into another miner's eyes. I tried to keep my light to the side of the coal miners Stan and I approached in the dark, while watching their lamps for signals. Moving your light in a circular motion means "approach." Moving the beam from side to side means "stop." Moving your light up and down in a nodding motion means "back up." Stan and I stopped at an intersection between tunnels.

The continuous miner, a coal cutter used before the longwall, had bored a hallway into a fresh seam of coal in the hallway adjacent to where we were. The roof bolter was arriving to shore up the new roof, a process that required six gauge steel mesh and long bolts driven into holes that were already stuffed with a tube of resin. The resin was mixed to form a stiff glue when the bolts twisted through the tube, breaking the seal between its two components.

Standing at the end of a hallway, in the dark except for the headlamps of the roof bolter and the lamps on our caps, Stan and I looked up, shining our lights into the meshed roof above us, where loose rock was held up only by the wire. "We don't have the best roofs," he murmured by way of explanation.

The rock base of the mountain was heterogeneous and liable to break apart. The mesh held up the cracked pieces of the roof, and the glue cemented the layers together. The roof bolter drilled holes, holding up the mesh with mechanical arms. One of the two operators driving the machine slid the resin in and placed the bolt into the machine. On my side of the hallway the wall shuddered, and coal sloughed off.

I jumped out of the way, and the last of the falling coal bruised my ankles through my rubber boots. A few minutes later, another layer slid off the wall, and I jumped again.

"You'd better stand over here," Stan offered, "or that'll keep bugging you. If you were standing against that there wall you coulda got hurt. A slab could slough off and smash you flat." We traded places, exchanging sides in the middle of the hallway,

Skyline Mine has received the Safest Mine Award from the Rocky Mountain Mining Association for two years running. Miners have weekly and monthly safety meetings. Their mascot is Bobcats, which also serves as an acronym: "Being Our Best Can Achieve Total Safety." Above the supply room is a flag depicting a mother holding a child, with the logo Sentinels of Safety, an award Skyline won in 2005 for being the safest mine in America, with twenty-eight months and five hundred thousand man-hours without any reportable injury. At the time I visited in July 2008, Skyline was celebrating two hundred fifty thousand man-hours without an accident. Nearly a year before, the roof bolter had malfunctioned and shot a bolt through a man's hand.

In 2006, the latest Bureau of Labor statistics ranked mining as the most dangerous occupation in America, with 49.5 fatalities for every one hundred thousand workers, outranking forestry, farming, and construction. The jump in mining deaths this year had a lot to do with the miners killed at Sago; in other years mining has come in second or third. In terms of injuries, it often ranks lower than these other hazardous industries, and this doesn't surprise me. The Occupational Safety and Health Agency (OSHA) is in charge of regulating the safety of all industries in America. MSHA is in charge of regulating only the mining industry, and it is staffed with far more inspectors per worker than OSHA.

As Stan talked about the safety precautions they have been required to implement, I remembered all the stunts we used to pull when I worked for the Forest Service. Once, we slid a one hundred pound generator down a steep ravine covered with willows, so we could finish a fence. Even on a daily basis, we swung axes to cut the ground free of sage brush without much training in how to use them properly. I would leave Skyline that day with one bruise, but during the two summers I worked for the Forest Service, my legs were so covered with bruises that a friend asked me if I was being abused. On one Forest Service project, the worker handling the chainsaw mis-cut a tree. The rest of the crew was relaxing on a log opposite to where the tree was supposed to fall. When the man operating the saw yelled, all my coworkers split to the sides. I ran straight, only outrunning the tree on the speed of adrenaline.

I understand the reluctance of the miners to follow their regulations and carry in that extra self-rescuer every working day. Our Forest Service crew leader, Emily, was always scolding us for taking off our hardhats or gloves. On most days it wouldn't have made a difference. One day, if Emily hadn't been wearing her hardhat, a barbwire fence-tightener would have taken off her head.

Mining is as safe as it is because of the regulations and their enforcement. Most mining accidents are the result of taking chances, but there is always that element of unpredictability. In West Virginia, they call them acts of God, but I might refer to them simply as acts of nature. Before we left the mine, Stan took me down a tunnel that was no longer being mined. We ducked under the tarp that blocked the entrance, and Stan shone his lamp up at the ceiling. I made out a dinosaur footprint—intact, with all three toes—embedded in the shattered rock of the roof.

"Arch wants it sent to headquarters in St. Louis for a gift."

Looking up at that footprint, I realized that being in the *bowels of the earth* meant dealing with powers that were supra-human because Arch Coal had no part in the laws that formed that print or the larger-than-life creature that made it. Miners call the walls of tunnels *ribs* instead of walls, because they know that being underground is like being in the belly of a prehistoric creature. There are layers built into the rock that an engineer can evaluate, but never dictate. The earth will do its own bidding, and there will always be factors beyond human control. This is something miners must respect, and something they must acknowledge each day when they step into a mine. Mining, by nature, is a hazardous occupation.

On our drive out, Stan delivered his last argument: "They call us 'evil.' Evil. The dirty fuel. They treat us like we're the plague. I'm not pretendin' that burning coal is better than natural gas because it isn't. But they make us out to be worse than we are. I just want them to tell the truth. It's not like it used to be."

Mining is *not like it used to be.* When I came out of the mine, I had coal dust smeared on my nose and across my cheeks, my ankle was throbbing, and my lungs hurt deep in my chest. But mining is not like it was, and I will never come close to approximating the experiences of my Welsh ancestors or the terror of my great-great-grandfather Evan when he descended down the mineshaft at the age of five. No miner in the United States ever will, and that is a credit to the industry and to progress in society as a whole. It is also worth noting that mining is not the same everywhere. There are better and worse ways to run a coal mine.

Since the Sago disaster , a coal mining accident occurred within twenty miles of Skyline. On August 6, 2007, the roof collapsed on six miners retreat mining in Crandall Canyon, trapping them 1,500 feet underground. The mining company Andalex had just been bought by Robert Murray, of Murray Energy, a company out of Cleveland, Ohio, with such a bad reputation in the East that nobody would give him a mining permit in Utah.

Soon after the deal was closed, Murray "bullied" a permit from MSHA to take out the pillars. The roof gave way, causing a small earthquake of 3.9 magnitude, which was recorded by University of Utah geologists. This is what's termed a *mine bounce*: the geological consequence of a mountain readjusting its weight. Rescue workers drilled from above while others started clearing the tunnel from below.

Fourteen miles away in Huntington, Utah, fifty family members waited at the local junior high school. Butcher paper signs hung all over town: "Pray for Our Miners"; "Proud to be an American, Proud to be a Miner"; "Dios Bendigas Nuestras Seis"—the locals dubbed the three Mexican and three American miners "our six." On the tenth day, three more were added to the list of casualties when a rib collapsed on three rescuers clearing the tunnel.

During the incident, Robert Murray appeared at press conferences, expressing concern in a baby blue sweater, a touch calculated to soften the impact of his heavy frame. On another occasion, he emerged to meet the reporters fully dressed as a miner, his face evenly blackened as if he'd smudged the coal on. "Had I known that this evil mountain, this alive mountain, would do what it did, I would never have sent the miners in here. I'll never go near that mountain again," he claimed. Throughout the news coverage of the incident, he came across as agitated, overly optimistic about rescue operations, and outraged at any indication that bad mining practices had been involved. I'll give him credit for genuinely hoping to recover his workers, but I also sensed his personal terror as the hours ticked past without any sign of life. At some point, the axe of justice will fall.

Crandall Canyon and one other Utah mine owned by Murray closed according to his word, but it would be better for the mining industry and miners if Robert Murray never went near any mountain again. The final report of the federal investigation into the Crandall Canyon disaster came out exactly two weeks after I visited Skyline. The federal government found MSHA negligent in ever issuing the permit, claiming it was "rubber stamped" without any serious oversight. The engineering firm, Agapito Associates, provided faulty recommendations for the mining design and was fined $220,000 by MSHA. A mining fine of $1.6 million, the heaviest fine in MSHA's thirty years of existence, was levied on the company. The US Senate Health, Education, Labor, and Pensions Committee recommended a criminal investigation, claiming that Murray showed "a callous disregard for the law and safety standards."

All mining pillars fail over time, but the majority fail relatively slowly. Moreover, seismologists cannot predict when mining pillars will fail. Every mountain is capricious—this is where nature comes in. But there are telltale signs. A rock burst happens when the ribs of a mine collapse because they are unable to absorb the pressure bearing down from the roof. In the section being mined by Huntington's "six," three rock bursts occurred before the mountain finally gave way. One of these rock bursts, in March 2007, prompted the mining superintendent to send this email to headquarters: "We've used all the tricks we know of to pull these pillars, and I no longer feel we can do it without unacceptable risk." The risks were sufficient enough that veteran miner Don Erickson recognized them and told his wife Nelda that the mountain shuddered violently when he worked inside. Three days before the fatal incident on August 6, another one of the ribs collapsed, and still the miners were sent in. Not only was the initial mining plan flawed (and on post-evaluation, the riskiest plan MSHA has ever approved), but the company also illegally removed sections of the pillars that were designated as off limits. After MSHA presented its report to the families of the deceased in Price, Utah, Representative George Miller, the chairman of the House Labor and Education Committee made some pointed comments. "MSHA's report affirms the conclusions reached by our own investigation: Murray Energy should not have proposed the flawed retreat mining plan, and MSHA should not have approved the plan. It is clear that Murray Energy is an outlaw company that recklessly endangered its employees' lives. It is tragic that the deaths of six miners and three rescuers resulted from the reckless actions of a few individuals and inadequate MSHA oversight."

Don Erickson was fifty years old, the father of two children, and a stepfather to three of his wife Nelda's children. His mother, Lucille, described him as "solid, loving, and considerate." Lucille and Erick Erickson lived in my parent's neighborhood in Price. "Our six" was something that all of Carbon County took personally. The city of Huntington erected a mural called "Heroes Among Us," depicting in relief the faces of the nine men who were killed. For six of the nine, it will be their only grave marker. Rescue efforts were officially disbanded by Congress shortly before the findings were released in July 2008. The miners were most likely killed by the mine bounce, and their bodies will never be recovered.

The reports on Crandall Canyon and Murray Energy shared headlines with Massey Energy Company. In 2007, the company earned record profits. In 2008, the company settled a lawsuit with the Environmental Protection Agency by paying a $20 million civil penalty for violating the Clean Water Act in its West Virginia and Kentucky coal mines. In July 2008, the company faced the possibility of losing half its estimated 2008 net income from a water pollution case in Rawl, West Virginia.

Massey's mountaintop removal operation is situated next to the town of Rawl, which lies below CEO Don Blankenship's mansion. Since Massey began mining there in 2003, the creeks have run black with coal sludge. Local townspeople have come down with rashes, and their hair has begun to fall out. Cancer, kidney and liver failure, and respiratory problems have increased exponentially over the past few years. In 2005, a study conducted by Wheeling Jesuit University revealed that the drinking water in Rawl's private wells exceeded federal limits for asenic, lead, iron, aluminum, beryllium, barium, manganese, and selenium—all chemicals typically found in coal sludge.

In the 1980s, Massey had pumped millions of gallons of coal sludge into underground cavities near Rawl. In the late nineties, when Massey prepared for its mountaintop removal operations, a blast cracked the foundations of dozens of homes. This is when the water began to go bad, and it is possible that the blast cracked the barrier between the underground sludge pond and Rawl's city aquifer. Massey refused to investigate the correlation or provide Rawl with safe drinking water. In 2004, seven hundred individuals filed suits against the company for the crimes of water pollution and manslaughter for two deaths caused by the contamination.

In 2008, Massey faced trial for the death of two miners in its underground Aracoma mine, which caught fire weeks after the Sago explosion. After pleading guilty to the safety violations that contributed to these deaths, Massey paid $2.5 million in criminal charges and $1.7 million in civil penalties, the largest plea bargain in the history of the coal industry. Widows Delorice Bragg and Freda Hatfield pleaded with the judge to reject the deal, holding Don Blakenship personally responsible for ignoring the memo sent to him six days before the fire, which clearly declaired the conveyor belt conditions unsafe.

In the same month the plea bargain for the Aracoma deaths was approved by Judge John Copenhaver, friends and family called to inform me of a devastating accident in another Massey mine. The Upper Big Branch coal mine in Montcoal, West Virginia, exploded on Monday, April 5, 2010, taking the lives of twenty-eight men and one woman—an astounding death toll for contemporary America. Montcoal is located on Coal River Road between Sundial and Sylvester, and press conferences reporting the status of the rescue efforts and the investigations into the causes of the accident were held at Marsh Fork Elementary School.

A spark from poorly maintained cutting machines lit methane gas, possibly escaping from underground reserves up through cracks in the mine floor. A massive accumulation of coal dust caught fire, and malfunctioning sprayers failed to extinguish the conflagaration. Five hours after the accident, seven miners were confirmed dead and nineteen missing. Two miners were receiving medical attention. Like the mining families four years previously at Sago Baptist Church, Romona and William Williams waited on the grounds of the Upper Big Branch mine for news of their brother. They talked to reporters through the window of their parked pickup truck as rescuers made slow progress into a mine where rails were twisted "into pretzels" from the heat and force of the explosion.

The next day, when Don Blakenship arrived at the mine to report the toll of twenty-five confirmed dead and four missing, he was escorted from the scene after the enraged crowd charged him with caring more about profit than human life. At the time, Blakenship had just turned sixty. The hair that had begun to recede from his temples had grayed. Lines of age and stress had softened the contours of his cheeks and jowls, but the lines around his mouth remained firm, accentuated by his neatly trimmed mustache.

Throughout Massey's history of enviromental and safety violations, Blakenship has always asserted his lack of culpability: "I think the fact that MSHA, the state, and our firebosses and the best engineers that you can find were all in and around this mine, and all believed it to be safe in the circumstances it was in, speaks for itself as far as any suspicion that the mine was improperly operated."

The accident came as a shock to the nation and to the federal government, which had implemented the 2006 MINER Act to prevent mining fatalities such as the ones at Upper Big Branch. But, as in the case of Crandall Canyon, there had been telltale signs. In the two months prior to the accident at Upper Big Branch, miners had been evacuated three times due to dangerous methane levels, and they had begun to feel nervous. Josh Napper, a twenty-five year old victim of the explosion, wrote his family a note the last day he went in: "If anything happens to me, I will be looking down from heaven." Clay Mullins, who lost his brother Rex, worked at Upper Big Branch until 2007 and asserted it had always been a gassy mine. He explained that when the older supervisors began to retire, their younger, less experienced replacements didn't know how to handle the high levels of methane.

The naturally ocurring methane was improperly controlled, but more condemning is the preventable issue of untreated coal dust. At Upper Big Branch, as at Castle Gate, a small fire became a blast that killed the miners inside almost instantly. Aside from servicing and replacing faulty equipment, simply employing rock dusting, the coal dust-neutralizing innovation that arose out of the Castle Gate disaster more than eighty years previously, could have saved twenty-nine lives.

As with Murray Energy, Massey Energy Company has a reputation for reckless mining. From March to April 2010, MSHA closed portions of three Massey-run mines due to safety violations. Inspectors, informed of illegal practices by anonymous tips, showed up at the mine sites unannounced and seized the phone lines, so mine operators couldn't tip off the men below. Hazardous conditions ranged from improper ventilation, to coal mining deep into roofs and walls beyond permitted areas, to water blocking escape routes. In November 2010, MSHA sought a federal court order to close the entire Freedom Energy mine in Pikeville, Kentucky, until the safety conditions improved—an unprecedented act in all of mining regulation history. According to M. Patricia Smith, the agency's top attorney, "This mine is basically an accident away from a possible tragedy."

A leaked memo that Blankenship sent to his deep mine superintendents in 2005 perhaps reveals the source of the company culture that leads to these failed inspections and dangerous mining conditions: "If any of you have been asked by your group presidents, your supervisors, engineers, or anyone else to do anything other than run coal, you need to ignore them and run coal ... This memo is necessary only because we seem not to understand that coal pays the bills."

George Miller's denunciation of Murray Energy as an "outlaw company" is paralleled by J. David McAteer's more measured criticism of Massey Energy Company. He called it "certainly one of the worst in the industry," with one of the "most difficult" safety records—a record for which Massey CEO Don Blankenship is yet to be held personally accountable.

• • •

There are better and worse ways to run a coal mine. Several months after the Crandall Canyon accident, I traveled to central Utah to meet with the former Vice President of Operations at Andalex. Sam Quigley had quit when Robert Murray signed the deal to take over the company. The contract went through at eleven o'clock in the morning, and Sam's office was cleaned out by noon. He was familiar with Murray's reputation and refused to work with him. Soon after, my dad recruited him to be his energy man at the College of Eastern Utah.

I asked Sam what he thought about the cave-in. "We know how to mine. We should be beyond that." At the time, he was reluctant to say anything about the case, as there was a possibility that he would have to testify in the national trials.

I visited Sam in the CEU energy center, located in the building that once served as a bath house for the Willow Creek Mine. The Castle Gate cemetery lies below, a reminder of the devastation possible in a mine. Throughout the center are signs typical of mining buildings: "Safety is a virtue," "Center for Behavior Based Safety," and "All Mining is Retreat Mining." The latter is Sam's personal motto, one that indicates the precautions a miner should take in any mining operation. There is always a chance the roof could come down.

A veteran miner, Sam is lean, wiry, and freckled. When I first met him, he caught me off guard with his fervor. After finishing a degree in geology at the University of Utah, Sam pursued the four-generation family business (metal mining) by working off his school loans digging a shaft for a trona mine in Green River, Wyoming. He wintered there in a school bus because there was a shortage of housing. Sam is not your typical miner, but he is miner's miner, emphatic in his defense of the trade.

Sam collects mining lamps and books. He drew diagrams to show me how the lamps worked and fired mining facts at me faster than I could scribble them down. In a way, he is a curiosity to me. A miner by choice, he worked during the years my family was absent and saw the transfer from the old way of mining a longwall to the new. In the early seventies, he worked in a coal mine in central Utah, putting up timbers (to hold up the roof) "with old time miners, the Palacio brothers, Pete, Manuel, and Johnny the Smoke." In those days, there were fourteen to fifteen mining companies in Carbon County, employing four thousand to five thousand miners who extracted eight million tons of coal. By 1990, there were only 1,700 left, but they mined around twenty-seven million tons.

Sam's response to the Crandall Canyon accident was more a commentary on how the public reacts to such incidents: "There's a whole group of people in society that believe that coal mining is an outdated activity and we don't need it anymore when nasty things happen. Your toothpaste has three mining products in it. Everything that you touch—this pen, which has plastic and metal—comes from mining."

Sam's sister was a teacher who was part of "that" society. When his boys went into mining, she tried to talk them out of it, telling them that "mining is no longer a necessary evil."

But Sam believed she had lost the historic vision, which he was adamant to share with me: "There was a time when the environment was a word that wasn't included in people's vocabulary. They were concerned about providing the basics. Miners are heroes. They lived in wood shacks at ten thousand feet during winters and provided the basis for everything that we have. Our society has lost sight of this. We are a blessed generation. We have the ability to protect our air and water. We can reclaim landscape. You know what gives us this ability? We have an affluent society. Affluence has given Americans the opportunity to think about the environment."

Like Sam's sister, I am a teacher, a third-generation teacher from four generations of coal miners before. It has been almost seventy years since any family member of mine has stepped into a coal mine for a day's wages. In sentiment and socialization, I belong to the society Sam lambasts, the one that believes—or wishes to believe—that coal mining is no longer a necessary evil. But coal mining is not inherently evil, and coal mining is necessary; it will continue to be so until Americans across the nation jointly decide to turn off their lights.

What Sam failed to say, I will say for him. Our affluence has been built by coal. We think about the environment because coal power provided the means to mechanize America. The engine was built in order to mine coal more efficiently, and it is on the power of this engine, derived from coal, that industrial America has become rich. Coal has given us the option of thinking about the environment, and it has given us the luxury of mining coal responsibly. This is where our task lies now, but not ultimately.

Coal is "the dirty fuel." There are health and environmental concerns with its procurement, processing, and burning, and there may be future concerns with the storage of its byproducts. To this day unsafe practices of major coal mine operators continue without regard for human life. This is current cost of our energy, but perhaps the source that provided the locomotion to move us through the twentieth century can pave the way for the next century to move beyond it.

24

The Last Deep Coal Mine in Wales

Early in January 2008, the Tower Colliery, the last deep mine in Wales, ceased operations. Until then, it was the oldest operating deep mine in the world. After eighteen years, the seams were exhausted, and the colliers who had fought so bravely against Margaret Thatcher were out of work. On hearing the news, I booked a flight to Wales. In all my exploration of coal and its history, coal has never gotten into my blood like my grandfathers before me, but coal miners, especially Welsh coal miners, plucked a particular ancestral strand. I wanted to get there before they filled in the shaft, so I could descend into the last coal mine in Wales. The United Kingdom still powers one-third of its electrical grid with coal, but jobs in the energy sector are on the decline. In the country that began the Industrial Revolution with steam power derived from its coal mines, coal mining is nearly a dead industry.

The Tower Colliery is just outside the town of Hirwaun, a name meaning "long meadow" in Welsh. A small town of four thousand, the streets curve through rows of quaint boxes of shops and homes. In our interview, Sam Quigley had suggested that part of coal mining's bad name came from the stereotype of the coal miner on a bar stool, but in this small Welsh town, I knew this was exactly the place to find them. I walked into the Glancynon, the town's most popular pub. It was early yet, but a group of gray-haired men lounged in a dark wooden booth across from the bar, watching curiously as I approached the bartender polishing glasses.

"I'm looking for colliers who worked in the Tower mine."

"Wayne there worked in the mine, now didn't ya?" She gestured toward a tall, shy, bearded man who stepped forward.

Wayne Joseph, an employee of the Glancynon, had mined for thirteen years and still lived in the coal miner's house he grew up in. When I asked him what he thought about the closing of the Tower, he answered right off, "I think it's a good thing, so people can move on with their lives."

He peered out at me from the corner of his eye to see if that was the answer I was looking for, but Wayne, in truth, had a different perspective from many of his comrades. He had left Tower several years before to work in the bar. He had never liked mining, considering it too dangerous.

"Why, then, do people stay?" I asked.

Wayne responded with the same answer miners had given me in West Virginia and in Utah. I began to wonder if it was the stock answer given by miners and agreed upon across continents: "Your grandfather, your father, it comes down the line. You could say it gets in a person's blood, basically." He added, "But you've got to think of the health. I'd say for every one hundred miners, eighty have problems with their lungs."

Alex, the cook at the Glancynon, arranged for me to meet with his father, Philip Edwards, the former environmental officer at Tower. Philip would meet me later that evening, so I took the opportunity to drive out to the mine. Just up the hill from Hirwaun and past the Penydarren brewery, I pulled over at a sign marked Tower Colliery and ducked under a metal fence. Before I got far, a blue minibus pulled up. A man wearing a neon colored hardhat got out and asked me what I was up to.

In Wales, I always told them that I was Erin Thomas and that my ancestors mined coal in Merthyr, hoping my last name and the local connection would get me somewhere. He introduced himself as Steve, and explained that I had ducked under the fence of the coal processing plant instead of the mine. Steve pointed at the belt transporting the coal and the series of pools. At Tower they cycled the water, cleaning it to be used again, rather than storing it in sludge ponds.

"Hop in, I'll take you to the mine," Steve offered, and swept off the front seat of a vehicle that used to be an airport shuttle. "It's a bit messy from the boys."

We drove not far down the road and pulled up where a gutting process was taking place. All the old equipment had been pulled up out of the mine, much of it dismembered to fit on the conveyor belt. Tires were piled around the periphery, and metal was heaped to be sold as scrap. It reminded me of the horses these machines had replaced, and how these horses had to be folded up to be brought in and out of the mine.

Steve led me to the shaft, warning me not to get too close. I had come to Wales too late to go down; the following week the mine shaft was scheduled to be filled in. A yellow strip of construction tape blocked the

Erin Thomas, 2008

Mine shaft at Tower Colliery

hallway to the elevator. The fans had been shut off, and I could smell the methane and feel the coal dust in my lungs. Standing at the mouth of that mine, I felt the same longing I had felt when I first entered Wales. I considered making a run for it just to look down.

At the sorting yard, coal was separated from the shale and then large stalls according to size: grains, beans, small, large. There were still massive piles of it. People from all over the world had come to buy this coal. Anthracite is the hardest grade of coal, and the hottest burning. Coal from Tower was only 4 percent ash, so pure that it had to be mixed with other materials before being sold to the power companies. Unless ash lumps, it is too hard to clean out of the bottom of broilers. Steve picked up a piece from the pile lovingly, "This coal is so clean."

Hours later, I sat down to drinks with Philip, his wife Mary, and his son Alex, who was taking a break from his cooking duties. Philip was a handsome man, with well-defined features and a charismatic bearing. His wife Mary had a shoulder-length bob that was just graying. They were all thin, but sturdy in the fashion of the Welsh. Mary and Alex leaned forward

Piles of coal at Tower Colliery

over the table as Philip related his story. Both Philip and Mary were of mining stock; Mary lost her brother in an explosion in 1959.

In the 1980s, when the miners went on strike, Philip found himself in the position of one of the ringleaders. He was named chairman of the Tower Colliery action committee and set off marching through the villages of the South Wales coalfields with a band and signs, rallying the people for support. He and his fellow miners went door to door for donations, food parcels, and sausage.

"Police on horseback charged the picket line, managed us with buttons."

"Buttons?"

"You know, 'buttons.'" Philip had told me if I didn't understand him to just "say," but it took a few tries before *baton* registered. I asked how they made ends meet while Philip was out of work.

He waved a hand and glanced at his wife, "Eighty-four through eighty-five was a bitter year. Mary worked part-time. Alex was two, and we got a milk token for 'em."

Mary nodded, "I worked part-time at the grocery store down the way."

After two years of striking, the miners went back to work until, one by one, each mine was closed. In 1995, they tried to close the Tower Colliery, but Philip and his buddies wouldn't budge. Officials offered the miners £10,000 each to walk away peacefully, but they wouldn't take it. The government then knocked their offer down to £9,000 to scare the miners into compliance, and the men were given two weeks to decide. Philip and his friends gathered at the Penywaun Welfare Club and formed the TEBO (Tower Employee Buy-Out) Team, taking their case to the legislature.

Philip scribbled on a piece of paper as he talked, writing down each consecutive price the government had offered them to walk away and crossing them out as the narrative progressed. "We were just a buncha coal boys. Two hundred and fifty of us pitched in £8,000 each. We took out a £3 million loan from the government, and paid it all back, every bit of it. We had 220 man in, and tonnage went up. In the end, we made £19 million.

"We were equal shareholders. We voted for how many holidays. Got a month off each year and sick leave. We had anniversary parties. One year, a bakery made a cake so big it filled the booty a' Mary's car."

Mary nodded, "The lady at the bakery painted a miner's cap on it."

From Philip's point of view, mining at Tower was like being in an adult playground. The men experienced the same sort of comradery that miners had relished for centuries, but they were able to take care of each other in a way a company or the government never would have.

"I had bowel cancer," Philip explained. "When I came back, the boys told me just to work as much as I could and fixed a toilet on the back of my trailer."

Philip and Mary invited me over to their home to look at pictures. A sports car was parked in the driveway. Philip led me through a backyard beautifully landscaped with a garden and trellises. We met Mary in their small front room next to the coal burning stove; she sorted through pictures of Philip leading the strikers during the eighties. Others were from later. Miners were posed in front of the colliery with their arms around each other, their hardhats on and their faces smudged. Throughout my research I had seen many pictures of miners. In older photographs, miners made straight faces, their carbide lamps clipped onto soft caps. Some of these men would have been paid in cash, and others with company scrip to spend in the company store. They would have lived in miner's houses rented from the company. Like Philip, some of them, at some point in their mining career, would have stood along a picket line disputing their share of the company's profits. These were Orwell's miners from the underworld, never talked about in polite society. Mining at Tower still had risks, and burning the coal dug there, despite its purity, still released carbon dioxide into the air. Although Tower recycled its water, there was still sludge to be dealt with, and the black piles of coal contrasted with the green grass of the hills. There were likely some among Philip's coworkers with black lung, but Philip's photos had a very different feel than many of the miners' photos I had seen. He pointed out his buddies, "George the Crane, Dai Milk the Bull, Dai Lumpy, and Squeaky." Squeaky, I gathered, was derived from an undesirable personal attribute that Philip was too polite to relate. They grouped together and grinned in front of a mine they owned, a triumph for all the Welsh coal miners before them.

As I left, Philip and Mary asked me to come again, but for all I knew it would be my last time in Wales. A Welsh coal miner was a dying breed, but I didn't need to worry about Philip and Mary. They would have plenty to do in their retirement. Only months before, they had visited their oldest son, who lived in Thailand and taught English, and they had plans to visit many other parts of the world.

Some of Philip's coworkers from the Tower Colliery would later be employed by Energybuild, a mining company investing in coal recovery from the old mining tips scattered throughout the hill of South Wales, the remnants of an industry that many Welsh believe defined their culture. Rhidian Davies, a veteran miner and now the company's managing director, would explain the venture: "Here is a saying which goes: once you have coal in your blood, unfortunately it stays there. We're just trying to keep an industry and a heritage alive." Villagers of Ellington, Lynemouth, and Cresswell would rise up in protest against the thirteen windmills to be installed in their communities, fearing it would prevent any hope of coal mining's comeback. And in Neath Valley, former Tower Colliery boss Tyrone O'Sullivan would counter climate change activists who opposed the opening of the mine owned by Unity Power (the first deep mine in sixty years) by saying: "There is no way to save the world. You can't stop burning coal because … you'd have no industrial base and no income being earned anywhere in the world."

How green was my valley then, and the valley of them that have gone. I left Wales in a harried drive from Merthyr to Cardiff to return the rental car, a brand new BMW I was sure I was going to crash when negotiating the roundabouts on the highway. I did not look at the hills as I had when I had first come, but kept my eyes alert for the small road signs hiding behind the wildflowers along the shoulder. I had three pieces of coal in my purse, sorted out by Steve from the pile marked "small." It was truly beautiful

Erin Thomas, 2008

Philip and Mary Edwards in front of their coal burning stove

coal, the type that the Romans had named jet and carved into jewelry. The coal gleamed black and was so hard, it barely rubbed away on my hands. Wayne Joseph said it burned "so fierce" that unless it was mixed with lower grades of coal, it would devour a fireplace in a week. I took one piece for my dad to remind him of his father, one piece for my grandfather's grave, and one piece to set next to my carbide lamp.

Holding the shiny piece of anthracite in my hands, I would think of Evan and Margaret, and Zephaniah and Maud. I would remember the little white chapel on Miner's Memorial Road, and wonder if Paul Avington had worked those last ten years in the mine. I would direct my thoughts toward Larry Gibson and the few days a year he spared from his mountain-saving crusade to enjoy his small house on top of Kayford Mountain, and toward Melvin Cook, hoping he had managed to hold onto his home in Blair. I would think of Stan Christianson, supervising the operations in the Skyline mine, and Sam Quigley and Philip Edwards adjusting to their lives aboveground. But most of all, this piece of anthracite, 96 percent carbon, would remind me daily of the price of my electricity. I would think of the intricate history and the current controversy involved in flipping a light switch. This pure Welsh coal would remind me to take direct responsibility for drawing from the grid.

25

The Winds of Change

At the turn of the century—we still refer to it as this (even though we have witnessed the turn of another) because this turn signified so much—humankind was for the first time experiencing mechanization. America was seized by the excitement of technology and recognized its source. Henry Obermeyer claimed: "Almost the sum total of modern construction in America is nothing more than a monument to coal."

As America surged to the lead in coal production, it surged to the lead in the world economy. It wasn't hard to remember where energy came from because at that time, a number of Americans still burned coal in their stoves. Evidence of the price of this energy loomed over industrial skies. In 1919, Waldo Frank wrote of Chicago: "The sky is a stain. The air is streaked with runnings of grease and smoke. Blanketing the prairie, this fall of filth like black snow—a storm that does not stop ... Chimneys stand over the world, and belch blackness on it. There is no sky now." In 1886, 31 percent of deaths from disease were smoke related: pneumonia, bronchitis, and asthma. Once viewed as a sign of progress, wealth, and job opportunities, smoke became seen as a degrader of health and morals.

Social change began with small suits brought by ordinary citizens against neighborhood polluters. In Pittsburg, one man sued a brick factory for damaging his orchards and grapevines, but as "manufacturing interests were necessary and indispensible to the growth and prosperity of every city," the Pennsylvania Supreme Court sided with the company. Tired of cleaning smoke residue off their curtains, antismoke women's organizations began to spring up in industrial cities across America. The Women's Health Protective Association of Allegheny County was formed in Pittsburgh; the Wednesday Club, in St. Louis; the Smoke Abatement

Club, in Cincinnati. These ladies combated smoke to promote the values of health, cleanliness, aesthetics, and morality, and they spent their afternoons patrolling residential skies for the smokestacks issuing the darkest fumes. Charles Reed, a member of the Smoke Abatement Club, declared before the Ladies Club of Cincinnati: "To breathe pure air must be reckoned among man's unalienable rights. No man had any more right to contaminate the air we breathe than he has to defile the water we drink. No man has any more right to throw soot in our parlors than he has to dump ashes into our bedrooms."

By 1916, seventy-five American cities had passed antismoke ordinances, and the nuisance suits brought against businesses that exceeded legal smoke density limitations were upheld in court. The Waterside electric plant along the East River in midtown Manhattan—owned by the upstanding American Thomas Edison—was fined for the smoke put out by its 114 broilers in People v. New York Edison 6. In order to abide by the new laws, companies appealed to the engineers. In this case, the answer was not necessarily the development of new technology, but priming existing equipment to minimize smoke output. There was no cure-all—each specific situation was assessed. With proper installation, settings, and operation of the equipment, plus "a good match between the equipment type and coal quality," factories and businesses reduced emissions. In some cases, this actually increased efficiency: low quality coal, when burned in a relatively smokeless manner, produced more heat. Skies began to clear, and became even clearer as Americans shifted to the fossil fuels of natural gas and oil, which were more efficient and cleaner burning.

This history of coal smoke eradication is a parable for our times, when we have more fully realized the atmospheric consequences of burning coal and other fossil fuels. It testifies to the validity of grassroots action to pressure energy companies into adopting pollution reducing measures. Smoke abatement verifies that persistent appeals for justice will convince courts that the health of citizens ultimately overrides any business concerns. Additionally, it provides evidence that our economy was able to sustain a shift in energy production. In our current situation, another production shift may be far overdue. The 105-year-old Waterside plant owned by Consolidated Edison (ConEd), fined for smoke output in 1916, was finally closed in 2001 as a result of pollution complaints by its neighbors. Its sister plant, the East River Generating Station in lower Manhattan, planned to pick up the slack by increasing operations until nearby residents living in what is dubbed "Asthma Alley" protested and sued the company. Mayor Rudolph Giuliani backed ConEd based on the familiar energy crisis rhetoric; the ensuing settlement required the company to pledge $3.7 million to clean up its operations. Despite the contributions of historical power plants to the industrialization of America, after a hundred years of pumping harmful emissions into our cities and our skies, it may be time to retire our grandfathers.

One of the most striking features of the smoke abatement movement is that the solution was embodied in the concepts of *specific* and *individual.* Just as engineers in the previous century were able to reduce the smoke output by evaluating particular broilers, coal grades, and atmospheric conditions, there are individual adjustments in energy use that nations, states, businesses and organizations, and citizens can make to reduce our carbon footprint and use of fossil fuels. Reclaiming our atmosphere will require a revolution of specific proportions to be enacted in multiple levels of society, which will pave the way for the use of alternative fuels and ultimately alternative energy lifestyles. This revolution may be initiated in part by citizens in Sylvester, Blair, and Alexandria who are tired of sweeping their porches—just as smoke abatement began with ladies tired of cleaning their curtains.

Although the US government has been sluggish about adopting nationwide energy regulations, thirty-two individual states and the District of Columbia have instituted renewable portfolio standards, mandating that a certain percentage of electricity be derived from renewable sources by a designated deadline. The particulars of each of these energy portfolios are specific to each state. California is one of the most ambitious, with a commitment to 20 percent of renewables by 2013 Among that state's possible alternative forms of energy are biomass, geothermal, hydropower, wind, and some of the most viable solar power capabilities in the nation. The Solar Initiative, which was launched on January 1, 2007, by Governor Arnold Schwarzenegger, provides funding to develop the capacity to produce three thousand megawatts of new, solar-produced electricity by 2017. Two years later, Schwarzenegger declared: "Our vision of solar panels lining the rooftops of houses and businesses across California is becoming a reality. I'm encouraged to see that even in these difficult financial times we are breaking solar installation records and spurring private investment in solar projects."

Individual cities are also initiating energy saving legislation. Pasadena, California, recently took up the challenge to become a model for green cities around the nation and the world. Adopting a environmental charter and endorsing the US Conference of Mayors' Climate Protection Agreement, which upholds the Kyoto Protocol, Pasadena has laid out an action plan that includes resolutions such as buying recycled paper, using green technologies in the city's transit system, buying products from local farms, and constructing buildings using energy efficient models and renewable sources of power.

Individual companies are also making efforts to build in more sustainable ways. During one of my visits to Utah, the opening of the Mark Miller

Toyota Dealership in Salt Lake City made headlines as one of the greenest buildings in the Intermountain West. It was built according to Leadership in Energy and Environmental Design (LEED) certification, with an emphasis on natural light, heat and energy saving technologies, waterless urinals, and internal components made of recycled materials. Although the initial construction of the building was costly, the green technologies cut its operational energy consumption by a third, and Miller expected payback from these efficiencies within ten years.

Aside from governmental and commercial initiatives, there are specific adjustments we can make personally in energy consumption. The Maynard family in Mesa, Arizona, keeps their air conditioning off until it is absolutely necessary, setting the temperature during the summer months at 80° F. They use a solar oven to cook bread, designate "no driving days," and coordinate errands on other days so that they don't have to use their vehicle frequently. They catch excess shower and sink runoff to water their plants. My friend Carl rides his bike to work every day and buys a share in alternative energy, so that even though in Virginia he only has access to traditional sources of power, he can help subsidize the higher price of alternative energy for somebody in another state. Laura and Dan in Seattle, Washington, buy into a farming cooperative and walk to the farmers' market every Saturday for their bundle of fruits and vegetables, saving the fossil fuel transportation costs of produce. In Utah, when my parents remodeled their home, they put in energy efficient light bulbs. My maternal grandfather bought a hybrid car and installed an energy efficient window shade in his living room; it blocks his view but cuts his air conditioning bills.

Transforming our atmosphere and averting an energy crises ultimately depend on our collective will for change. With new technologies for coal burning and carbon capture, the easy answer may be to continue burning the remaining recoverable coal reserves and let a future generation to deal with the environmental and human health costs of further extraction and the implementation of alternative technologies. Symbolically, our energy future is the decision between the two pillars that marked my entrance into Wales and the beginning of my exploration into coal—a smokestack and a windmill.

I was reminded of this image when traveling through the mountains to visit the Skyline Mine in central Utah. Nine wind turbines turned almost imperceptibly at the mouth of Spanish Fork Canyon. Tall, white, industrial-chic, they did not blend well with the Rockies. But they gave me hope. I could smell the outside air pumped in through the air conditioning. It was particulate; a "red" day the announcer had warned on the radio, not a good day to be outside. It was fortunate I was headed underground.

In the eye-sore of those windmill blades, there were no CO_2 emissions, no coalcleaning stations, no sludge ponds, no mines. They did not blend well, but in their turning there was a chance of something cleaner. Utah lags behind other states, with only one megawatt of installed wind power. In the state's energy portfolio, the goal of 20 percent alternative energy production by 2025 is not mandatory, but a geological survey of the prime areas for wind power has been prepared by the US Department of Energy, and the Utah Wind Working Group has been founded.

This is why, when my father hired Sam Quigley to run the energy center at the College of Eastern Utah, he hired him to train energy workers, and not coal miners. Although Sam is quick to assert that "coal is the only resource that works," he also acknowledges times are changing, coal is an exhaustible resource, and the 1,100 miners who work in thirteen Utah mines may be doing something else in the future. Classes at the center focus on the areas of mechanics, electrical systems, hydraulics, equipment operation, and ground control—all transferable skills allowing energy workers to shift between technologies. In 2007, it was estimated that 50 percent of coal miners and 50 percent of coal plant workers would be eligible for retirement in the next decade. Coal mining especially is struggling to find new recruits, and the market presents a natural opportunity for jobs in new energy generation technologies, which have finally become feasible and reasonably affordable.

In my current home state of Virginia, wind is one of the viable alternatives. It receives a double credit under the voluntary energy portfolio goal that grants incentives to private businesses to generate energy from alternative sources. Although there are no commercial wind farms in Virginia to date, the neighboring states of West Virginia and Pennsylvania have both installed profitable wind facilities.

Especially in West Virginia, wind power may be a partial solution to fill the hole in the economy that scaling back on coal mining would inevitably create. Citizens and environmental groups have realized that in order to save their mountains, they must replace mountaintop removal with another source of income and energy that is less destructive. The Coal River Mountain Watch, an environmental organization based in West Virginia, has proposed the Coal River Wind Project, a plan to erect a wind farm on Coal River Mountain in Raleigh County, West Virginia, which is currently permitted to Massey Energy for a mountaintop removal project. Not only would the project preserve the mountain and the forest, but it would also provide more income for the community and the state.

According to a study conducted by Downstream Strategies, this wind farm will permanently generate $1,740,000 in county taxes, in contrast with the strip mine, which will only pay $36,000 over a period of just seventeen years. Jobs will be available to local residents indefinitely, while the coal mining jobs will only last for the duration of the strip mine. Downstream Strategies also calculated that the health (in terms of death and hospitalization) and environmental costs of mountaintop removal ultimately exceed the financial benefit to the local community from employment and taxes. In another town in West Virginia, Dominion Resources and BP are currently investigating the possibility of installing a sixty megawatt generator with thirty turbines, which will provide sufficient energy for fifteen thousand homes. The companies anticipate this project will make a profit of $2 million annually, offer five to ten permanent jobs, and provide several thousand dollars in taxes each year.

In response to Pennsylvanian consumer requests, the Green Mountain Wind Farm was the first wind facility to be installed in the small borough of Garrett in 2000. National Wind Power, a UK company, erected the eight, two hundred foot tall turbines on land that was reclaimed from a coal strip mine and is currently farmed by the Decker family. When it was built, it was one of the largest wind farms on the East Coast; since then, Pennsylvania has erected six wind farms, one in the neighboring town of Meyersdale. Green Mountain Wind Farm generates twenty-five million kilowatt hours each year, powering 2,500 homes.

Early in April, when the trees where just beginning to sprout leaves, I traveled to Pennsylvania to inspect the current most feasible alternative to coal. Garrett and Meyersdale are former coal mining communities nestled in hills. The topography did not rise sufficiently to meet my classification of mountain country, but the land rippled into a contour that made me feel at home. As in Appalachia, turning off a major highway plunged me into a stream of small towns: the borough of Salisbury, the village of Boyton. The residences were a mixture of beat-up sheds, trailer homes, older wooden and brick houses, and newer edifices with aluminum-siding sheens. As in small-town West Virginia, these Pennsylvanians expressed their patriotism by decorating their houses with Americana. There were fewer beaten-brass stars; these Yanks proclaimed their pride of being over the Mason-Dixon Line with personal flagpoles that spanned divergent local economies. I spied a flag waving next to a large stone establishment, and a flag waving on the property of a ramshackle white trailer.

Near Green Mountain, in valleys that sloped down from the hills, wide-roofed red barns lounged next to silver grain silos—seeming to suggest farmland wholesomeness and prosperity. The straw colored land that rolled out in front was squared by barbed wire nailed to wooden dancers. Small communities clumped between the stretches of field. In Meyersdale, the houses climbed to the ridge of a hill that was topped with a line of white turbines built three years after the Green Mountain facility.

I gasped when I reached the summit of the next hill and the windmills popped up on the horizon. Although the wind farms hadn't seemed to jibe with the local scenery in Utah, either the clean lines were beginning to grow on me or turbines resonated in a state I associated with the

Erin Thomas, 2009

White turbines in Meyersdale, Pennsylvania

Pennsylvania Dutch; calmness and self-sufficiency seemed to emanate from that line along the hilltop. The slow turn of the massive, ninety-five foot blades felt meditative—a measured spin that said what must be done must be done thoughtfully. I skirted the edge of one property along a line of pussy willows to get a closer photograph of the windmills next to a pine. A man and his wife were selling Easter eggs on the side of the highway for a church fundraiser, and I asked them what they thought of the facility. They shrugged their shoulders and handed me my change.

Past Meyersdale, you come to the edge of Garrett, a town founded in 1871 by its namesake, Robert T. Garrett, the son of a railroad man. According to the 2000 census, 449 people live there. Just on the outskirts, the turbines of Green Mountain are visible above the Cassleman River and the railroad track that runs alongside, an artery that carried Pennsylvania coal to energy consumers in the outside world. The rock that once powered Garrett's economy still had local appeal. On entering the town, I immediately caught sight of a little white sign advertising "House Coal." Piles of wood were stacked behind most of the homes, and I realized a good portion of the residents burned both coal and wood for heat.

I stopped to talk to a tall man in his forties with buzz-cut hair, who was fixing the sign of the local Baptist church. He was dressed in scrubby jeans and a fleece jacket, and I initially assumed he was a repairman, but given the size of the population, he may well have been the preacher. I asked him his opinion of the wind farm: "First an attraction, now a distraction. It'd be nice if we benefited from it, but they pipe all the energy out." When the facility was first installed, it attracted thousands of visitors and wind energy experts. The dedication was marked by a celebration that included live music, drinks, and a barbecue, an event that may have provided temporary employment—and certainly entertainment—for the surrounding towns. The keynote speaker, Dan Reicher, then the Assistant Secretary for Energy Efficiency and Renewable Energy, declared, "Today we are celebrating a new opportunity for Pennsylvania and throughout rural America."

Like the MTR operations in West Virginia, the energy produced by Green Mountain Wind Farm doesn't power the homes of the locals, but is carried elsewhere. An extractive business, it's operated by international CEOs who are the primary financial beneficiaries. It's a shame that Green Mountain doesn't provide electricity to the town of Garrett, given the amount of energy that is lost over transmission lines. Despite Reicher's comments, wind is probably not the ultimate solution to unemployment or poverty in rural America. In 2008, wind power made up one-third of new electrical generation, creating ten thousand energy jobs. This is nothing to the percentage of the population coal employed at the peak of bituminous extraction in 1923 when 705,000 men worked underground. For that matter, current mining doesn't compare favorably, either. In 2007 there were eighty-one thousand coal miners—for a new technology, wind is not far behind.

Compared with the grievances in Appalachia over coal dust, explosions, floods, and poisoning, most residents in Garrett had little to say. Aesthetics is typically the major complaint against the installation of new wind farms. Once facilities are in operation, surrounding communities have voiced concerns of noise pollution and lowering property values. In Nantucket, Maryland, since 2001, local citizens have been protesting the installation of the first offshore wind farm. The proposed 130 turbines would provide 420 megawatts of pollution free power, but opponents counter that the "steel forest" would ruin the scenery and interfere with fishing.

In my interview with Jim Presswood, I had asked him for his opinion about the situation: "If you're going to build a turbine within sight of million-dollar homes, you're going to have problems. You can develop offshore a farther distance out beyond the horizon. Sighting is an issue, and we're very sensitive to those concerns. Wind turbines on top of mountain ridges for instance, that's not necessarily the thing to do, but think of a mountain that's been leveled; maybe that makes more sense."

Robert Kennedy Jr., who spoke at the MTR gathering I attended in Kayford, West Virginia, was very vocal in the debate against wind farms along the northeast Atlantic coast. Aside from disrupting breeding grounds for sea creatures, concerns have also been raised about seafaring migratory birds. There is no truly "green" approach to energy, but a visit to Garrett, Pennsylvania—after my trip through coal country in Wales, after walking the path of Zeph and Evan in central Utah, after witnessing the modern reality of mining in West Virginia, and understanding the true impact of coal burning plants on my lungs—was a detoxifying experience.

Coal had begun get to under my skin, like the blue scars on the faces of my Welsh ancestors. As an energy consumer, I was complicit in this process of coal. It was under my fingernails, smudged on the end of my nose, and no matter how many times I scrubbed, the coal dust would not come free of my clothing. It had worked its way into everything that I used and ate; coal dust was laced through the internal organs of my body.

The energy that ran in the wires out of Garrett was free from the stain of coal sludge and toxic heavy metals. It meant healthier air and streams that ran clearer. Garrett would never be obliterated from the map by dynamite and draglines. The lives of these citizens in small-town Pennsylvania would remain intact. The turn of these windmills constituted a streamlined process of energy generation to the power lines, no dig, no cleanse, no burn—straight from the blade to the grid. As my coal miner's lamp and the anthracite lump from Tower Colliery will always remind me of the coal miner's blood that runs in my veins, these turbines will symbolize a flush and a freedom from the rock that has shaped my life.

Afterword

The arc of the moral universe is long, but it bends towards justice.
—Martin Luther King Jr.

One Saturday in November, I looked up from an ultimate frisbee game in Anacostia Park to spot a bald eagle. The park lies upstream from Old Town Alexandria, where the Potomac River splits and straddles the downtown area of the nation's capital.

"The power plant along the river developed a bald eagle nesting program to boost their environmental creds," I told a friend. "Maybe that's one of their birds."

Five days after the eagle sighting on Thanksgiving Day, I had the worst asthma attack I've experienced in seven years after playing twenty minutes of ultimate frisbee. I called for two friends to carry me off the field after a half an hour of trying to walk back home had left me twenty feet from the sidelines of the game, sitting in a mud puddle. Three days later, I parked a block from the Potomac River Generating Station, now owned by GenOn, a company created by the merger of Mirant and RRI Energy in April 2010. I scuttled along the railroad track with my camera, looking for a good shot and hoping that nobody would confront me for snooping.

The five squat smokestacks were barely visible from my worm's-eye view, and airplanes zoomed overhead every few minutes. I climbed a small hill opposite the plant and focused my lens. I leaned against a fence that separated the railroad track from an apartment complex. The switch from residential to industrial felt too sudden; townhouses lay within fifty feet of the plant's open pile of coal. As I took pictures, bikers and runners passed on the path that curves around the plant along the river. I had taken pictures when I visited the plant before, but it seemed appropriate to return—on August 30, 2011, GenOn agreed with the City of Alexandria to close the Potomac River plant by October 1, 2012, signaling the end of an era.

With the Pepco lines in, the DOE no longer deemed the plant necessary for the capital's electrical redundancy. The City of Alexandria will surrender the $32 million held in escrow for environmental improvements after the 2008 settlement and has pledged to help GenOn find work for the 120 plant employees who will lose their jobs.

Even before GenOn announced the plant's closure, the nonprofit American Clean Skies Foundation proposed plans for redeveloping the twenty-five-acre site. The venture, named Potomac River Green, includes office, retail, hotel, and residential space, as well as several unique features: a Clean Energy Enterprise Center, an Energy Museum of twenty-first-century alternatives, and a compressed natural gas and electrical car refueling station. The bikers and runners will still enjoy recreational space, with an eco-trail extending into nearby Dangerfield Island. Plans include reusing the plant's brick and steel in the rebuilding process.

In response to an online *Washington NBC* article announcing the closure, Sandra Wheeler wrote,

> So happy to hear the news—the coal fired Pepco facility is closing its doors permanently in Alexandria! I was diagnosed with COPD [chronic obstructive pulmonary disease] after living a few blocks from this facility for 10 years! Oddly, since moving, no more issues with breathing, even with living in allergy burdened Eastern North Carolina. Doctors here have declared no COPD.

The most common cause of chronic obstructive pulmonary disease is smoking, but pollution and frequent exposure to smoke are also common causes. There is no documented cure. During 2010, the Potomac River Generating Station emitted 1.1 million tons of CO_2, 1,145 tons of NO_x, and 1,400 tons of SO_2. In spring 2011, the State of Virginia fined the plant $275,000 in civil penalties for emissions violations. The plant's closure is truly a triumph for the City of Alexandria and the health of its residents. On learning the news, I felt amazed and delighted, but I also experienced a tinge of regret. Despite my own struggles with asthma, I couldn't help feeling sorry for Lynwood Reid, the kind operations team manager who had served as my guide on the plant tour. A city took on a corporation and won, but Alexandria is a city of affluence. GenOn will continue to operate, and 120 blue-collar energy workers will struggle to find employment in a depressed economy.

Rarely do company executives pay the price for our energy. Public opinion is set against Robert Murray and Don Blankenship, but as of this writing no material punishment has been levied for their involvement in the some of the greatest mining tragedies of our time.

In July 2010, four foremen of the Massy owned Aracoma Mine received criminal convictions for their safety violation of failing to conduct escapeway drills. This might have prevented the deaths of the two miners who got lost in the fire. In response to a web article on the topic in the *Charleston (WV) Gazette*, several readers made comments. Don wrote, "Sad to say this, but it's probable that they were all doing what they needed to keep their jobs. Somehow, the real culprits never get called to account for their misdeeds." Monty followed, "Reading through all of these stories, one gets the impression that while these men were doing things they probably knew they shouldn't have been doing, they were also sacrificial lambs." Daniel R. Cobb finished with, "Fines? Only fines? Does anyone ever go to prison?"

In October 2011, Hughie Stover, the safety director at the Upper Big Branch mine, the Massey operation where twenty-nine miners died in April 2010, was convicted of lying to federal investigators and ordering the destruction of records. He denied that Massey had a policy of tipping off foremen when federal investigators arrived for surprise inspections, a testimony challenged by that of mine guards who delivered the tips through an established and sophisticated system of radios, walkie-talkies, and strobe lights. For this offense, Stover could serve up to twenty-five years in prison.

Five months later, mining superintendent Gary May was charged with conspiracy to defraud the government, which carries a potential sentence of five years in prison. May's subversive activities included rigging the ventilation system to trick safety officials and disabling a methane monitor on a cutting machine. On February 23, 2012, the day after May was charged, West Virginia Office of Miners Health Safety and Training released its report on the Upper Big Branch mine disaster, citing 253 violations for which it could fine the company up to nearly a million dollars in penalties. The state investigation upheld that the cause of the explosion was an improperly maintained cutting machine, which sparked a small pocket of methane gas, which ignited an excess build up of coal dust, which clogged and malfunctioning water sprayers were unable to extinguish. News reports have hinted at May's intentions to plea bargain for a lower sentence in exchange for information. US Attorney Booth Goodwin in charge of inspections indicated his probe was "absolutely not" finished, possibly indicating his aims go beyond May's prosecution to former Massey executives.

May is the highest level mining employee ever charged in a criminal case, but the conviction of a fellow miner, however culpable, is not what the families of the victims of the Upper Big Branch explosion and the Aracoma fire are really after. Aracoma widows Delorice Bragg and Freda Hatfield want to see Blankenship behind bars. After the foremen were convicted in the Aracoma case, the widows' lawyer stood and read a statement: "US Attorney Goodwind must tell Congress now ... why the law is insufficient to pursue all persons responsible for such blatantly criminal and deadly conditions." Patty Quarles, whose son Gary died in the Upper Big Branch mine, expressed a similar sentiment, "Somebody called the shots, and I don't think these section bosses and mine foremen can be held accountable alone. It goes higher than that. It goes all the way up the ladder to Blankenship ... He needs to pay for what he's done." After West Virginia released its report on the Upper Big Branch accident,

Jack Bowden, father-in-law of deceased miner Steve Harrah, expressed his relief that the criminal investigation was still progressing. He had no qualms about sharing his hopes for its outcome, "It was murder, basically. Don Blankenship should be held accountable."

Don Blankenship had a reputation for micromanaging his mines, and at Upper Big Branch in particular, he required miners to fax him production reports every half hour. Blankenship has a long record in West Virginia for underhanded and reckless mining practices, but the Upper Big Branch disaster opened him up to wider scrutiny. In December 2010, *Rolling Stone* published an article dubbing Blankenship "the Dark Lord of Coal Country." A week later, reportedly from pressure from the firm's board of directors, Blankenship retired and disappeared from public view. His generous retirement package included twelve million dollars in cash, insurance, an office and a secretary, access to Massey's documents in case of future litigation, and a 1965 blue Chevrolet. In June 2011, Massey Energy shareholders approved the company's sale to Alpha Natural Resources.

The fate of Bob Murray and Murray Energy still hangs in the balance. Criminal investigations by the US Attorney's Office of Utah into the Crandall Canyon disaster drag on, which may make it more difficult to accurately assess blame. J. David McAteer, former assistant secretary of MSHA, expressed his opinion, backed by years of experience in mining safety: "It's a shame, just a damned shame … Unfortunately, it suggests that in the end, it will peter out. The longer these things go, the more likely they are to be resolved without prosecution. That's an accepted fact in criminal work or civil matters."

Investigations focus heavily on the general manager of Crandall Canyon, Laine Adair, and his possible concealment and false representation of hazardous conditions. Representative George Miller, senior Democrat on the House Education and Workforce Committee, wrote the following to his fellow committee members: "Adair and others at [Murray Energy] may have purposefully misled MSHA about the severity of the March bump fearing MSHA would close the mine, and [they] continued to adhere to the mischaracterization after the August incidents." Miller's reference to "others" may very well hint at Robert Murray, owner of Murray Energy.

• • •

The arc of the moral universe is long, but it bends towards justice. On the fortieth anniversary of Martin Luther King Jr.'s assassination, then Senator Barack Obama added to King's statement: "It bends towards justice, but here is the thing: it does not bend on its own. It bends because each of us in our own ways put our hand on that arc and we bend it in the direction of justice." Recent events confirm this assertion, and not only in cities like Alexandria, Virginia. Persistent grassroots action and the courage of private citizens to take their grievances against corporations to the courts have begun to bear fruit in many small towns in rural West Virginia.

In 2009, three years after Ed Wiley's march to Washington, D.C. for the Pennies for Promise Campaign, Senator Byrd finally took action, publically criticizing Massey for refusing to donate funds to rebuild Marsh Fork Elementary School in Sundial, West Virginia. In March 2010, Don Blankenship folded, promising $1 million to the $10.7 million project (Alpha Natural Resources increased the donation to $1.5 million). Coal River Mountain Watch, a local environmental organization, donated $10,000. In April 2010, Charles Annenberg Weingarten of the Annenberg Foundation visited Appalachia when he heard about the Upper Big Branch explosion, learned of the children's precarious situation, and donated $2.5 million from his family's organization to supplement other private donations and the funds from West Virginia's School Building Authority and the Raleigh County School Board.

The following is taken from a website announcing the Annenberg donation: "Led by the residents of the Coal River Valley, supported by environmental, community, and human rights groups and celebrities, the 'powers that be' were no match against dedicated and persistent people. WAY TO GO EVERYONE. When the history of the movement against mountaintop removal is written, the victory at Marsh Fork will be remembered as a key moment. Celebrate ... celebrate ... power to the people!"

On April 30, 2010, an informal coalition of environmentalists and community members gathered at Ed Wiley's house for "roasted pork butt and wild turkey" served with potato salad. A year and a half later, on October 6, 2011, two hundred elementary school children wearing yellow hardhats sang at a groundbreaking ceremony attended by their parents and various dignitaries. Three miles from the current location of Marsh Fork Elementary School, construction is set to finish in December 2012 on a new, energy-efficient, red brick facility.

After seven years of litigation over the toxic slurry that leaked into city aquifers in Rawl, West Virginia, 769 people received compensation for gastrointestinal disorders, cancers, and learning disorders caused by lead and other heavy metal poisoning. The 350 lawsuits were settled on July 27, 2011, in a confidential hearing at the Kanawha County Courthouse; soon after, the figure of thirty-five million dollars was leaked to the press.

The families of Jackie Weaver, Marshall Winans, Tom Anderson, Jesse Jones, Jerry Groves, and George "Junior" Hamner received wrongful death settlements on November 16, 2011, for the mining explosion at Sago. Six other families had already received compensation, and Martin Toler's family declined to sue. ICG, the company operating Sago Mine at the time of the disaster, has been acquired by Arch Coal. Lawyer Al Karlin, who represented four of the families, expressed the following:

"This is a somber moment. When a case like this is settled, it's too easy to forget what it was all about. No settlement can really resolve what the families are left with—the holes in their lives."

In spite of safety legislation prompted by the Sago Mine disaster, more egregious safety violations contributed to the deaths of a total of forty individuals in the Aracoma, Crandall Canyon, and Upper Big Branch mining disasters. Karlin continued, "In some ways [safety conditions] are better today, but in some ways we're still facing the problems of safety in coal mines."

In West Virginia and across the US, the names Massey Energy and ICG will forever be associated with catastrophe. Now under the ownership of Alpha Natural Resources and Arch Coal respectively, perhaps these mines will use the distance from these names to make a distinct departure from the past in their company values. Immediately after buying out Massey, Alpha Natural Resources launched a "Running Right" safety program, training 7,500 miners. On December 6, 2011, the company agreed to pay $210 million in a settlement with federal prosecutors for the Upper Big Branch Mine accident, including payments to the families for wrongful deaths, money for mining safety research, and funds to clear Massey's unresolved fines with MSHA. This protects Alpha Natural Resources from further liability as a *corporation*, but it doesn't protect *individuals* associated with the cause of the explosion from criminal prosecution.

Despite appropriate financial compensation, there is one man with whom coal miners and citizens in many rural towns in West Virginia have yet to settle a score.

Alpha's settlement for the Upper Big Branch explosion once more drew the media's view toward Don Blankenship, and journalists have uncovered an alarming development. Only a month after stepping down as CEO of Massey Energy, Blankenship filled out paperwork for the McCoy Coal Group Inc. of Belfry, Kentucky. The company, bearing the maiden name of Blankenship's mother, has yet to seek a permit, but its establishment signals his only too clear intention of making a return to the mining industry. I find it noteworthy that Blankenship has picked a location in Appalachia, where mining laws are least enforced, for his comeback.

Phil Smith, the brawny and articulate spokesman for the UMWA, did not mince words on hearing the news: "Don Blankenship belongs in jail, not in a position to put yet more miners' lives at risk." In 2011, after concluding its investigations of the Upper Big Branch disaster, the UMWA designated Massey "a rogue corporation" and its actions, "industrial homicide." MSHA likewise determined that the responsibility for the Upper Big Branch explosion was a direct result of Massey's "systematic, intentional, and aggressive efforts" to hide unsafe working conditions—to the extent that managers kept two sets of inspection books, an accurate one for the mine and a falsified one for regulators.

The Federal Coal Mine Safety Act of 1952, the Federal Coal Mine Health and Safety Act of 1969, the Federal Mine Safety and Health Act of 1977, and the Mine Improvement and New Emergency Response Act of 2006 have all extended the power of the federal government to protect miners and to take legal action against mines that perpetuate unsafe working conditions. The families who lost their loved ones in the Aracoma, Upper Big Branch, and Crandall Canyon mines now wait for federal investigators to take the next historic step: holding a company executive personally responsible for the institutionalized disregard for human life.

Acknowledgments

This book would never have been if it weren't for Cornel and Gretchen Thomas, who lovingly compiled a binder of our Thomas family history. Special thanks go to family members Ryan and Ann Thomas, Patrick Thomas, Megan Hebdon, Mike Thomas, and Caitlin Thomas for their willingness to read draft after draft, and to Kyoko Mori for her advice and guidance in developing the first half of this manuscript for my master's thesis. Gratitude is extended toward Amanda Dambrink for editing my book proposal and providing commentary on several chapters, and Jared Gillins for supplying a number of sources and saving the day by reminding me to back up my manuscript days before my computer crashed. And then to Zach Winter, who helped me recover some of the lost documents. Fond remembrance to Sandie and John Robinson and Anna Headington for their graciousness in helping me with housing and offering other assistance in Merthyr Tydfil. A salute goes out to Abby Falls and Greg Smith, who made pleasant companions on my research trips, and to all my friends who sent me articles or otherwise supported me throughout the writing process. I am especially grateful for the peer reviews provided by Philip Notarianni, Allan Kent Powell, Kenny King, Carolyn Jacob, and Rick Bradford.

Sources

Due to the narrative nonfiction style of *Coal in Our Veins: A Personal Journey*, I have not used footnotes throughout the text. Interviews were provided by Carolyn Jacob, Emylyn Davis, Malcolm, Anna and Nathan Headington, Wease Day, Paul Avington, David K. Minturn, Chuck Nelson, Ralph Cook, Melvin Cook, Glenda Smith, Larry Gibson, Ann Carter, Sam Quigley, Stan Christianson, Phil Smith, Janet Gellici, Jim Presswood, Greg Jackson, Wayne Joseph, and Philip Edwards. The genealogical information included in this volume was gathered from texts written by Mary Jane Thomas Jones, Cornel W. Thomas, Margaret Wilding Thomas, Janet Thomas, Shirley W. Thomas, and Robert K. Thomas. The research compiled for this book was extensive. Consequently, only a selected bibliography has been provided, which is organized topically. Sources with duplicate information or sources consulted for minor details have not been listed.

Merthyr Tydfil and Welsh Mining History

Anonymous. "A Miner's Catechism." In *Coal: An Anthology of Mining*, edited by Tony Curtis, 168–170. Bridgend, Wales: Seren, 1997.

Borrow, George Henry. *Wild Wales: Its People, Language, and Scenery*. New ed. New York: Scribner and Welford, 1888.

Cannon, Mike. "Festival Celebrates Welsh Heritage." *LDS Church News*, March 13, 1993.

Davies, Herbert Edmund, Chairman. *Report of the Tribunal Appointed to Inquire into the Disaster at Aberfan on October 21st 1966*. London: Her Majesty's Stationery Office, 1967.

Done, Stephen, and Claire Dovey-Evans. *Cyfartha Castle Museum and Art Gallery*. Merthyr Tydfil, Wales: Merthyr Tydfil County Borough Council Leisure Services Department, 2000.

Evans, R. Meurig. *Children in the Mines 1840–42*. Cardiff: National Museum of Wales, 1979.

Gross, Joseph. "Merthyr Tydfil: The Development of an Urban Community." *Merthyr Historian* 5 (1992): 59–75.

———. "Water Supply and Sewerage in Merthyr Tydfil 1850–1974." *Merthyr Historian* 2 (1978): 67–78.

Hughes, John L. "Nothing Special." In *Coal: An Anthology of Mining*, edited by Tony Curtis, 44–45. Bridgend, Wales: Seren, 1997.

Jones, Bill, and Beth Thomas. *Coal's Domain: Historical Glimpses of Life in the Welsh Coalfields*. Cardiff: National Museum of Wales, 1993.

Llewellyn, Richard. *How Green Was My Valley*. New York: Macmillan, 1940.

"Local Intelligence." *Merthyr Express*, June 22, 1870.

Morris, J. H., and L. J. Williams. *The South Wales Coal Industry, 1841–1875*. Cardiff: University of Wales Press, 1958.

Orwell, George. *The Road to Wigan Pier*. New York: Harcourt, Brace, 1958.

Owen, David. *South Wales Collieries*. Vol. 4 of *Images of South Wales*. Gloucestershire: Tempus, 2003.

Paget, William, and Mary Paget. *Man of the Valleys: The Recollections of a South Wales Miner*. Gloucester: A. Sutton, 1985.

Rapoport, I. C. "Aberfan: The Days After; A Journey in Pictures." Cardigan: Parthian, in association with the National Library of Wales, 2005.

Sikes, Wirt. *Rambles and Studies in Old South Wales*. London: S. Low, 1881. Quoted in "Merthyr Tydfil." Genuki.org. http://www.genuki.org.uk/big/wal/GLA/MerthyrTydfil/.

Spartacus International. "Penydarren." http://www.spartacus.schoolnet.co.uk/RApenydarren.htm.

Trevelyan, Marie. *Glimpses of Welsh Life and Character*. London: J. Hogg, 1893.

Vaughan, F. "Some Aspects of Life in Merthyr Tydfil in the Nineteenth Century." *Merthyr Historian* 3 (1980): 87–97.

Watkins, Hugh. "The Story of Martyr Tudful." *Merthyr Historian* 3 (1980): 9–16.

Welsh Mormon History. http://welshmormonhistory.org/.

Williams, Huw. "Brass Bands, Jazz Bands, Choirs: Aspects of Music in Merthyr." *Merthyr Historian* 3 (1980): 100–110.

Scofield Mine Disaster

"The Death Toll Had Reached 250." *Salt Lake Tribune*, May 4, 1900.

Dilley, J. W. *History of the Scofield Mine Disaster: A Concise Account of the Incidents and Scenes That Took Place at Scofield, Utah, May 1, 1900, When Mine Number Four Exploded, Killing 200 Men*. Provo, UT: Skelton, 1900.

Israelsen, Brent. "Mine Disaster in Carbon County Won't Be Forgotten." *Salt Lake Tribune*, May 31, 1999.

Mele, Kimberly. "Part II. 'One Spark of Time': The Winter Quarters Mine Disaster." *Holmes Safety Association*, July 2007. http://www.holmessafety.org/2000/july00.pdf.

"Most Appalling Mine Horror!" *Salt Lake Tribune*, May 2, 1900.

Norris, Leslie. "Elegy: For the Men Killed at Winter Quarters, Scofield, Utah." Provo, UT: Tryst Press, 2000.

Reid, Robert N. *The Scofield Mine Disaster*. Lincoln, NE: BR Enterprises, 2002.

Utah's Castle Country. "Winter Quarters Coal Mine Disaster." http//:www.castle-country.com/what_to_do/mining_winter_quarters.html.

Watt, Ronald G. "Explosion of Pleasant Valley Coal Company." Utah History to Go. http://historytogo.utah.gov/utah_chapters/statehood_and_the_progressive_era/explosionofpleasantvalleycoalcompany.html.

Castle Gate

Campbell, Marius R. *Guidebook of the Western United States: Part E. The Denver & Rio Grande Western Route.* Washington, DC: United States Printing Office, 1922.

Costa, Janeen Arnold. "A Struggle for Survival and Identity: Families in the Aftermath of the Castle Gate Mine Disaster." *Utah Historical Quarterly* 56 (1988): 279–292.

Craig, Williard. Interview by Mark Hutchings. *Oral History Project of the Charles Redd Center for Western Studies at BYU.* June 12, 1976.

Halverson, Eugene. "Castle Gate." OnlineUtah. http://www.onlineutah.com/castlegatehistory.shtml.

Hartley, Jo. "Former Negro Preacher Breaks Down Prejudice." Utah Genweb Project, August 24, 1997. http://www.carbon-utgenweb.com/history.html.

"J. M. Burns, Castle Gate Officer, Slain by an Infuriated Negro." *Sun Advocate* [Price, UT], June 19, 1925.

Lofthouse Publishing. "Castle Gate Mine Explosion." http://www.lofthouse.com/USA/Utah/carbon/castlegatex.htm (site discontinued).

"Marshall Taken by Crowd and Swung Up." *Sun Advocate* [Price, UT], June 19, 1925.

Notarianni, Philip F. "Hetacomb at Castle Gate, Utah, March 8, 1924." *Utah Historical Quarterly* 70 (2002): 63–74.

Reeve, W. Paul, and Jeffery D. Nichols. "Klansmen at a Funeral and a Terrible Lynching." *History Blazer: News of Utah's Past,* September 1995.

Thacker, Faye. Interview by Mark Hutchings. *Oral History Project of the Charles Redd Center for Western Studies at BYU.* September 20, 1976.

"Utah Mine Blast Buries 175 Men." *Salt Lake Tribune,* March 9, 1924.

Carbon and Emery County Mining History

"Abandoned Building Dynamited: Main Objective; 'To Kill a Ghost.'" *Sun Advocate* [Price, UT], July 10, 1969.

Arch Coal, Inc. "Arch Coal's Skyline Mine Achieves Nation's Best Underground Safety Record, Earns Sentinels of Safety Award." September, 22, 2006. http://news.archcoal.com/phoenix.zhtml?c=107109&p=irol-newsArticle&ID=1360764&highlight=.

Ashby, Kevin. "Boylen Details Progress at Willow Creek." *Sun Advocate* [Price, UT], June 24, 1999.

Cunningham, Frances. *Driving Tour Guide: Selected Abandoned Mine Sites; Castle Coal Country, Carbon and Emery Counties.* Helper, UT: Peczuh, 1990.

Powell, Allan Kent. *The Next Time We Strike: Labor in Utah's Coal Fields, 1900–1933.* Logan, UT: Utah State University Press, 1985.

Taniguchi, Nancy J. *Castle Valley America: Hard Land, Hard-Won Home.* Logan, UT: Utah State University Press, 2004.

Taylor, Troy. *Down in the Darkness: The Shadowy History of America's Haunted Mines, Tunnels & Caverns.* Alton, IL: Whitechapel Productions Press, 2003.

Whitley, Colleen, ed. *The History of Mining in Utah.* Logan, UT: Utah State University Press, 2006.

US Coal-Mining History

Alinsky, Saul. *John L. Lewis: An Unauthorized Biography.* New York: G. P. Putnam's Sons, 1949.

Arrington, Leonard J., and John R. Alley. *Harold F. Silver: Western Inventor, Businessman, and Civic Leader.* Logan, UT: Utah State University Press, 1992.

Browning, Albert D. *Death, Destruction, and Disaster in the American Coal Mining Industry.* Bloomington, IN: 1st Books Library, 2003.

Coleman, Ron. *We Dig Coal: The Story of Coal Mining in Buchanan County, Virginia.* Redford, VA: Commonwealth Press, 1975.

"Death at Donora." *Time Magazine,* July 3, 2009. http://www.time.com/time/magazine/article/0,9171,853334,00.html.

Dubofsky, Melvyn, and Warren Van Tine. *John L. Lewis: A Biography.* Abridged Version. Urbana: University of Illinois Press, 1986.

Energy Information Administration. *Longwall Mining.* Office of Coal, Nuclear, Electric, and Alternative Fuels, US Department of Energy, March 1995. http://www.eia.gov/ftproot/coal/tr0588.pdf.

———. *US Coal Industry, 1970–1990: Two Decades of Change.* Washington, DC: Office of Coal, Nuclear, Electric, and Alternative Fuels, US Department of Energy, 2002.

Fishback, Price V. *Soft Coal, Hard Choices: The Economic Welfare of Bituminous Coal Miners, 1890–1930.* New York: Oxford University Press, 1992.

Francaviglia, Richard V. *Hard Places: Reading the Landscape of America's Historic Mining Districts.* Iowa City: University of Iowa Press, 1991.

Levy, Elizabeth, Tad Richards, and Henry E. F. Gordillo. *Struggle and Lose, Struggle and Win: The United Mine Workers.* New York: Four Winds Press, 1977.

Lockard, Duane. *Coal: A Memoir and Critique.* Charlottesville: University Press of Virginia, 1998.

National Mining Association. "Trends in US Coal Mining 1920–2007." http://www.nma.org/pdf/c_trends_mining.pdf.

Ratay, Desiree Southern. "The Orient No. 2 Coal Mine Explosion." *The Boneyard: A Journal of History, Opinions, Adventures, and Ideas.* http://www.evansville.net/user/boneyard/orient.htm.

Rogers, William P. *The Coal Primer.* Van Buren, AR: Valley Press, 1978.

General Mining History

Hayes, Geoffrey. *Coal Mining.* Buckinghamshire, UK: Shire, 2004.

Pohs, Henry A. *Early Underground Mine Lamps: Mine Lighting from Antiquity to Arizona.* Tucson: Arizona Historical Society, 1974.

Thurston, Robert H. *A History of the Growth of the Steam-Engine.* New York: D. Appleton, 1878.

United Mine Workers of America. "Black Lung." http://www.umwa.org/?q=content/black-lung.

Coal as an Energy Source

Environmental Integrity Project. "50 Dirtiest Power Plants." July 27, 2009. http://www.environmentalintegrity.org/pub385.cfm (site discontinued).

Folger, Tim. "Can Coal Come Clean? How to Survive the Return of the World's Dirtiest Fossil Fuel." *Discover*, December 2006, 42–48.

Gabbard, W. Alex. "Coal Combustion: Nuclear Resource or Danger." *Oak Ridge National Laboratory* 26 (Summer/Fall 1993). http://www.ornl.gov/info/ornlreview/rev26-34/text/colmain.html.

Goodell, Jeff. *Big Coal: The Dirty Secret Behind America's Energy Future.* Boston: Mariner Books, 2006.

Gwynne, S. C. "Coal Hard Facts." *Texas Monthly*, January 2007.

Hawkins, David G., Daniel A. Lashof, and Robert H. Williams. "What to Do About Coal." *Scientific American* 295 (September 2006): 69–75.

Martin-Amouroux, Jean-Marie. "Coal: The Metamorphosis of an Industry." *International Journal of Energy Sector Management* 2, no. 2 (2008): 162–180. http://proquest.umi.com (access restricted).

Thwaites, Tim. "The Mine of the Future." *Earthmatters: CSIRO Exploration and Mining Quarterly Magazine*, September 2003, 3–5.

US Geological Survey. "Health Impacts of Coal Combustion." July 2000. http://pubs.usgs.gov/fs/fs94-00/fs094-00.pdf.

US House of Representatives. *Carbon Capture and Sequestration: Hearing before the Subcommittee on Energy and Air Quality of the Committee on Energy and Commerce.* Washington, DC: US Government Printing Office, 2008.

———. *The Future of Fossil Fuels: Geological and Terrestrial Sequestration of Carbon Dioxide; Joint Oversight Hearing before the Subcommittee on Energy and Mineral Resources, Joint with the Subcommittee on National Parks, Forests, and Public Lands of the Committee on Natural Resources.* Washington, DC: US Government Printing Office, 2007.

US Senate. *Clean Coal, Oil, and Gas Development: New Energy Opportunities through Carbon Capture and Storage; Hearing before a Subcommittee of the Committee on Appropriations.* Washington, DC: US Government Printing Office, 2007.

———. *Clean Coal Technology: Hearing before the Committee on Energy and Natural Resources.* Washington, DC: US Government Printing Office, 2007.

Alternative Energy

Arvizu, Dan E. "Integrating Innovation and Policy for a Renewable Energy Future." Presented at the Kennedy School of Government at Harvard University, February 5, 2007. National Renewable Energy Laboratory. http://www.nrel.gov/docs/fy07osti//41166.pdf.

City of Pasadena. "Pasadena's Green City Homepage." http://www.cityofpasadena.net/GreenCity/.

Dougherty, P. "Toward a 20% Wind Electricity Supply in the United States." Presented at the 2007 European Wind Energy Conference, Milan, Italy, May 7–10, 2007. Washington, DC: National Renewable Energy Laboratory, 2007 [preprint].

Downstream Technologies. "Executive Summary: The Long-Term Economic Benefits of Wind versus Mountaintop Removal Coal on Coal River Mountain, West Virginia." December 2008. Coal River Mountain Watch. http://www.coalriverwind.org/wpcontent/uploads/2008/12/wind-executive-summary.pdf (site discontinued).

Geothermal Resources Assessment Team. "Assessment of Moderate and High Temperature Geothermal Resources of the United States." US Geological Survey, 2008. http://pubs.usgs.gov/fs/2008/3082.

Giauque, Marc. "Dealership Built with Environment in Mind." KSL, June 25, 2006. http://www.ksl.com/index.php?nid=148&sid=3614234.

"Green Mountain Wind Farm Dedicated." *Weekend Adventures Magazine*, Fall 2000. http://www.wamonline.com/fall2000/greenmountain.htm.

Hand, Maureen, Nate Blair, Mark Bolinger, Ryan Wiser, and Rich O'Connell. "Power System Modeling of 20% Wind-Generated Electricity by 2030." Presented at the IEEE Power and Energy Society 2008 General Meeting, Pittsburgh, Pennsylvania, July 20–24, 2008. New York: IEEE [Institute of Electrical and Electronics Engineers] Power and Energy Society, 2008.

Harrison, K. W., and G. D. Martin. "Renewable Hydrogen: Integration, Validation, and Demonstration." Presented at the 2008 NHA Annual Hydrogen Conference, Sacramento, California, March 30–April 3, 2008. Washington, DC: National Hydrogen Association, 2008.

Lovins, Amory. Interviewed by Scott Bilby. "Dr. Amory Lovins Talks about Energy Efficiency, Transport and Renewable Energy." *Beyond Zero*, February 26, 2009. http://www.beyondzeroemissions.org/amory-lovins-rocky-mountain-institute-energy-efficiency/.

Moore, Patrick. "Going Nuclear: A Green Makes the Case." *Washington Post*, April 16, 2006.

"Resisting Wind Power: Maryland Is Among the States Resistant to New, Alternative Energy Options." *Plenty Magazine*, March 24, 2009. http://www.mnn.com/earth-matters/energy/stories/resisting-wind-power.

Sullivan, Patrick, Walter Short, and Nate Blair. "Modeling the Benefits of Storage Technologies to Wind Power." Presented at the American Wind Energy Association (AWEA) WindPower 2008 Conference, Houston, Texas, June 1–4, 2008. Washington, DC: American Wind Energy Association, 2008.

Unkefer, Charlie. "The Bright Side: Local Solar Company Booms." *Mt. Shasta Area Newspapers*, February 11, 2009. http://www.mtshastanews.com/news/x1848777646/The-bright-side-local-solar-company-booms.

US Department of Energy. "Report to Congress on Assessment of Potential Impact of Concentrating in Solar Power for Electricity Generation." Energy Efficiency and Renewable Energy, February, 2007. http://www.nrel.gov/docs/fy07osti/41233.pdf.

———. "States with Renewable Portfolio Standards." EERE [Energy Efficiency and Renewable Energy] State Activities & Partnerships. http://apps1.eere.energy.gov/states/maps/renewable_portfolio_states.cfm.

———. "Virginia Incentives/Policies for Renewables and Efficiency." DSIRE [Database of State Incentives for Renewables & Efficiency]. http://www.dsireusa.org/library/includes/incentive2.cfm?Incentive_Code=VA10R&state=VA&CurrentPageID=1&RE=1&EE=1.

US Senate. *Electricity Generation: Hearing before the Committee on Energy and Natural Resources.* Washington, DC: US Government Printing Office, 2008.

West Virginia Coal-Mining History

Montrie, Chad. *To Save the Land and People: A History of Opposition to Surface Coal Mining in Appalachia.* Chapel Hill: University of North Carolina Press, 2003.

Shogan, Robert. *The Battle of Blair Mountain.* Boulder, CO: Westview Press, 2004.

Shuford, Chuck. "What Happens When You Don't Own the Land." Daily Yonder: Keep it Rural, July 3, 2009. http://www.dailyyonder.com/what-happens-when-you-dont-own-land/2009/07/03/2205.

Tams, W. P. *The Smokeless Coal Fields of West Virginia.* Morgantown: West Virginia University Press, 2001.

West Virginia Geological and Economic Survey. "History of West Virginia Mineral Industries—Coal." http://www.wvgs.wvnet.edu/www/geology/geoldvco.htm.

West Virginia Office of Miners' Health, Safety and Training. "West Virginia Coal Mining Facts." February 12, 2007. http://www.wvminesafety.org/wvcoalfacts.htm.

Sago Mine Explosion

Associated Press. "Official Apologizes to Families in Mine Tragedy." MSNBC.com, January 4, 2006. http://www.msnbc.msn.com/id/10682163.

———. "Text of Sago Mine Survivor's Letter." *Seattle Times,* April 28, 2006. http://seattletimes.nwsource.com/html/nationworld/2002957935_minelet-ter28.html?syndication=rss.

———. "2 Ex-Sago Miners Commit Suicide." CBS News, September 27, 2006. http://www.cbsnews.com/stories/2006/09/26/national/main2042071.shtml.

Grose, E. R. *History of Sago Community.* Morgantown, WV: Agricultural Extension Division, 1926.

Harki, Gary. "Questions Linger after Release of Sago Report." *Exponent Telegram* [Clarksburg, WV], July 23, 2006.

Harlan, Chico. "Sago Mine: Shredded Skin, Scarred Psyches for Disaster's Survivors." *Pittsburgh Post-Gazette,* October 8, 2006. http://www.postgazette.com/pg/06281/728326-357.stm.

Heyman, Daniel. "Mine Union's Report on the Sago Disaster Contradicts Earlier Findings." *New York Times,* March 16, 2007. http://www.nytimes.com/2007/03/16/us/16sago.html.

Huber, Tim. "Sago Mine Where 12 Died Is Idled in W.V." *Casper (WY) Star Tribune,* March 21, 2007. http://www.highbeam.com/doc/1Y1-104467115.html.

Langfitt, Frank. "A Bitter Saga: The Sago Mine Disaster." NPR, January 7, 2006. http://www.npr.org/templates/story/story.php?storyId=5134307.

Majors, Dan. "Coal Town Struggles to Move On a Year after Sago Mine Disaster." *Pittsburgh Post-Gazette,* December 24, 2006. http://www.postgazette.com/pg/06358/748704-357.stm.

"The Miners Trapped in the Coal Mine." Post by Missy1970, April 14, 2006. Violition.com Forums. http://forum.volition.com/topic.asp?TOPIC_ID=43872.

Norris, Michelle. "Pastor Recalls Time with Miners' Families." NPR, January 4, 2006. http://www.npr.org/templates/story/story.php?storyId=5126624.

"Profiles: Tallmansville Miners." CNN, January 8, 2006. http://www.cnn.com/2006/US/01/08/miners.bios/index.html (article discontinued).
Schoonover, Kelly. "Families Quietly Mark Sago Mine Disaster." Star News Online, January 3, 2007. http://www.firerescue1.com/carbon-monoxide-poisoning/articles/274371-Families-quietly-mark-Sago-mine-disaster/.
"Sound of Moans Led Rescuers to Surviving Miner." CNN, January 5, 2006. http://www.cnn.com/2006/US/01/04/mine.explosion.wed/index.html.
Ward, Ken, Jr. "Judge Approves Sago Lawsuit Settlements." *Charleston (WV) Gazette*, November 16, 2011. http://wvgazette.com/News/201111160248?page=2&build=cache.
———. "Memorial Honors Sago Miners." *West Virginia Gazette*, August 21, 2006. http://wvgazette.com/News/TheSagoMineDisaster/200608210003.
———. "MSHA Citations Detail Sago Problems." *West Virginia Gazette*, June 3, 2007. http://www.highbeam.com/doc/1P2-14770331.html.
Winans, Helen. Interview by Alex Chadwick. "A Miner's Mother on Risks, Tragic Loss." *Day to Day*, NPR, January 4, 2006. http://www.npr.org/templates/story/story.php?storyId=5126097.

Mountaintop Removal and Other Abuses

Broder, John M. "Rule To Expand Mountaintop Coal Mining." *Washington Post*, August 22, 2007.
Danawv. "Victory: New School for Marsh Fork Elementary." It's Getting Hot in Here: Dispatches from the Youth Climate Movement, April 30, 2010. http://itsgettinghotinhere.org/2010/04/30/victory-new-school-for-marsh-fork-elementary/.
Eilperin, Juliet. "EPA to Scrutinize Permits for Mountaintop-Removal Mining." *Washington Post*, March 25 2009. http://www.washingtonpost.com/wp-dyn/content/article/2009/03/24/AR2009032401607.html.
Fahrenthold, David A. "Mountaintop Mining to Get More Scrutiny." *Washington Post*, June 11, 2009. http://www.washingtonpost.com/wp-dyn/content/article/2009/06/11/AR2009061102205.html?hpid=moreheadlines.
Jervy, Ben. "Marsh Fork Elementary School Will Escape the Mass of Toxic Coal Sludge." Good Environment, May 12, 2010. http://www.good.is/post/marsh-fork-elementary-school-will-escape-the-mass-of-toxic-coal-sludge/.
Liptak, Adam. "Motion Ties W. Virginia Justice to Coal Executive." *New York Times*, January 15, 2008. http://www.nytimes.com/2008/01/15/us/15court.html?ei=74b7724453256d61&.
Loeb, Penny. "Sheer Madness." *US News and World Report*, August 11, 1997, 28–36.
"Marsh Fork Breaks Ground on New School." WVVA.com, October 6, 2011. http://www.wvva.com/story/15634049/marsh-fork-breaks-ground-on-new-school.
Mitchell, John G. "When Mountains Move." *National Geographic*, March 2006, 104–123.
Nyden, Paul J. "All Sides Mum at Rawl Settlement Hearing." *Charleston (WV) Gazette*, September 29, 2011. http://www.wvgazette.com/News/201109292579?page=2&build=cache.
Peterson, Erica. "All We Wanted Was Water." West Virginia Public Broadcasting, January 26, 2009. http://www.wvpubcast.org/newsarticle.aspx?id=7742.

Reece, Erik. *Lost Mountain: A Year in the Vanishing Wilderness; Radical Strip Mining and the Devastation of Appalachia.* New York: Riverhead Books, 2006.
Shnayerson, Michael. *Coal River: How a Few Brave Americans Took On a Powerful Company and the Federal Government—to Save the Land They Love.* New York: Farrar, Straus and Giroux, 2008.
Shogren, Elizabeth. "Bush Administration Altered Appalachian Landscape." NPR, April 15, 2009. http://www.npr.org/templates/story/story.php?storyId=99493874.
Simone, Samira. "Tennessee Sludge Spill Runs over Homes, Water." CNN.com, December 24, 2008. http://www.cnn.com/2008/US/12/23/tennessee.sludge.spill.
Ward, Ken, Jr. "Memo Could Be Key in Aracoma Case." *West Virginia Gazette,* April 17, 2010. http://wvgazette.com/News/montcoal/201004170245.
Warrick, Joby. "Appalachia Is Paying Price for White House Rule Change." *Washington Post,* August 17, 2004. http://pqasb.pqarchiver.com/washington-post/access/679433211.html?FMT=ABS&FMTS=ABS:FT&date=Aug+17%2C+2004&author=Joby+Warrick&pub=The+Washington+Post&edition=&startpage=A.01&desc=Appalachia+Is+Paying+Price+for+White+House+Rule+Change
White, Davin. "Massey Changes Course, Pledges $1 Million to Marsh Fork School." *Charleston (WV) Gazette,* March 23, 2010. http://www.wvgazette.com/Life/darcieboschee/mostRecent/201003230420 (article discontinued).

Potomac River Generating Station

"Alexandria Power Plant to Close." *NBC Washington,* August 30, 2011. http://www.nbcwashington.com/news/local/DC-Alexandria-Power-Plant-to-Close-128674048.html.
Burnely, Robert G. Letter to Lisa D. Johnson, August 19, 2006. Alexandria.gov. http://alexandriava.gov/uploadedFiles/tes/info/91305 Update to the City Council on Mirant.pdf.
Bertele, Alice, personal communication. Letter, September 24, 2007.
Bright Ideas Alexandria. "About the Hearing." May 22, 2007.
City of Alexandria. "Mirant Potomac River Power Plant: Air Quality and Health Concerns." September 13, 2005. Alexandria.gov. http://alexandriava.gov/uploadedFiles/tes/info/91305 Update to the City Council on Mirant.pdf.
Clean Air Task Force. "Health Impacts of Air Pollution from Washington DC Area Power Plants." May 2002. http://www.catf.us/resources/publications/files/DC_Metro_Summary.pdf.
Commonwealth of Virginia Department of Environmental Quality. "Stationary Source Permit to Operate." Alexandria.gov. http://alexandriava.gov/uploadedFiles/tes/oeq/info/MirantPRGSFinalPermit.pdf.
de la Merced, Michael J. "Merger of Energy Producers to Form $3 Billion Company." *New York Times,* April 12, 2010. http://query.nytimes.com/gst/fullpage.html?res=9C07EFDB1031F931A25757C0A9669D8B63&ref=mirantc orporation.
Downey, Kirstin. "Longtime Foes Face Off Over Mirant Power Plant." *Washington Post,* December 18, 2007. http://pqasb.pqarchiver.com/washingtonpost/access/1400190021.html?FMT=ABS&FMTS=ABS:F T&date=Dec+18%2C+2007&author=Kirstin+Downey+-+Washington

+Post+Staff+Writer&pub=The+Washington+Post&edition=&startpage=B.1&desc=Longtime+Foes+Face+Off+Over+Mirant+Power+Plant%3B+N.Va.+Dispute+Puts+Ex-US+Officials+at+Odds+Again.

Durrer, Austin. "Moran Statement on Alexandria Settlement with Mirant Power Plant." Congressman Jim Moran, July 2, 2008. http://moran.house.gov/list/press/vs08_moran080702.shmtl (site discontinued).

GenOn. "Potomac River Generating Station". http://www.genon.com/company/stations/potomac.aspx.

Hunt, Ralf M, personal communication. Letter, January 22, 2008.

Krouse, Sarah. "Details of GenOn and Alexandria's Deal to Close Mirant Plant." *Washington Business Journal,* August 30, 2011. http://www.bizjournals.com/washington/blog/2011/08/details-of-genon-and-alexandrias-deal.html.

Levy, Jonathan. "Executive Summary: Analysis of Particulate Matter Impacts for the City of Alexandria, VA." April 2004. Alexandria.gov. http://alexandriava.gov/uploadedFiles/tes/info/Dr.%20Levy's%20Study.pdf.

Neibauer, Michael. "No Power Issues from Alexandria Plant Closure." *Washington Business Journal,* September 30, 2011. http://www.bizjournals.com/washington/blog/2011/09/no-power-issues-from-alexandria-plant.html.

Shogran, Elizabeth. "Despite Health Risks, D.C. Power Plant Kept Open." NPR, August 20, 2006. http://www.npr.org/templates/story/story.php?storyID=5673425.

Smith, Leef. "Alexandria Pushes to Shut Mirant: Officials Skeptical of Cleanup." *Washington Post,* March 21, 2006. http://pqasb.pqarchiver.com/washingtonpost/access/1006714111.html?FMT=ABS&FMTS=ABS:FT&date=Mar+21%2C+2006&author=Leef+Smith&pub=The+Washington+Post&edition=&startpage=B.02&desc=Alexandria+Pushes+to+Shut+Mirant%3B+Officials+Skeptical+of+Cleanup%3B+Executive+Says+Plant+Has+the+%27Right+to+Run%27.

Tower Colliery

Black, Dave. "Lynemouth Villagers Are Set for Wind Farm Misery." *Journal* [Newcastle upon Tyne, UK], January 10, 2009. http://www.journallive.co.uk/north-east-news/todays-news/2009/01/10/lynemouth-villagers-are-set-for-wind-farm-misery-61634-22661287/.

Devine, Darren. "'You Cannot Run the World Without Coal': Ex-Colliery Leader Hits Out at Greens' Claims over Future of Fuel Supplies." *South Wales Echo,* March 20, 2009. http://www.walesonline.co.uk/news/wales-news/2009/03/20/you-cannot-run-the-world-without-coal-91466-23188294/.

Heatherington, Peter. "Pitting Their Wits." *Guardian* [Manchester, UK], February 20 2006. http://www.guardian.co.uk/society/2008/feb/20/communities.welshcollieries.

"Miner-Owned Pit Is Facing Closure." BBC, January 27, 2007. http://news.bbc.co.uk/2/hi/uk_news/wales/4653710.stm.

"The Miners' Strike." BBC Wales History, http://www.bbc.co.uk/wales/history/sites/themes/society/industry_coal06.shtml.

Webb, Tim. "Welsh Miner Finds a Welcome in the City: Rhidian Davies Was On Strike for a Year as a Miner; As Managing Director of Energybuild, He Wants to Preserve a Welsh Way of Life." *Observer* [London], March

15, 2009. http://www.guardian.co.uk/business/2009/mar/15/rhidian-davies-energybuild-mining.

Crandall Canyon

Buckley, Cara, and Susan Saulny. "Finding Miners Alive Is 'Totally Unlikely,' Owner Says." *New York Times*, August 23, 2007. http://www.nytimes.com/2007/08/23/us/23mine.html?adxnnl=1&pagewanted=print&adxnnlx=1325140077-pSQL52CkS1Sg8OTBpHtHyg.

Frosch, Dan. "Reports on Mine Collapse Criticize Operation and Oversight." *New York Times*, July 26, 2008. http://www.nytimes.com/2008/07/26/us/26mine.html?_r=1&adxnnl=1&oref=slogin&pagewanted=print&adxnnlx=1217106143-apIvZrgwZfjRa9gVyL9GsQ.

Gehrke, Robert. "MSHA Concedes It Could've Done More." *Salt Lake Tribune*, July 26, 2008. http://www.sltrib.com/News/ci_10004595.

"Lawmakers React to Report." *Salt Lake Tribune*, July 24, 2008. http://www.sltrib.com/news/ci_9986653.

Smart, Christopher, and Nate Carlisle. "Hope Fades for Quick Rescue." *Salt Lake Tribune*, August 7, 2007. http://www.sltrib.com/ci_6565193.

Upper Big Branch Mine

Davidson, Samuel. "Security Official Convicted of Obstructing Probe into West Virginia Mine Disaster." World Socialist Web Site, November 2, 2011. http://www.wsws.org/articles/2011/nov2011/ubbm-n02.shtml.

Harki, Gary A. "Miner Mourns Lost Brother, Friends." *Charleston (WV) Gazette*, September 25, 2010. http://www.wvgazette.com/News/montcoal/201009250986.

James, Michelle. "A Timeline Look at the Upper Big Branch Mine Disaster." *Register Herald* (Beckley, WV), April 11, 2010. http://www.register-herald.com/local/x1612533014/A-timeline-look-at-the-Upper-Big-Branch-Mine-disaster.

Lillis, Mike. "MSHA Closed Three Massey Mines in Last Month." Washington Independent, April 27, 2010. http://washingtonindependent.com/83331/msha-closed-three-masseymines-in-last-month.

Smith, Vicki. "Upper Big Branch Mine Conditions Were 'Industrial Homicide,' Union Alleges." *Huffington Post*, October, 25 2011. http://www.huffingtonpost.com/2011/10/25/united-mine-workers-massey-energy_n_1030542.html.

———. "Final report on W. Va. mine blast comes amid charge." *Deseret News*, February 23, 2012. http://www.deseretnews.com/article/765553369/Final-report-on-WVa-mine-blast-comes-amid-charge.html.

"Some Justice at Upper Big Branch." *New York Times*, October 27, 2011. http://www.nytimes.com/2011/10/28/opinion/some-justice-at-upper-big-branch.html.

"Upper Big Branch Mine Disaster—Special Online Section." *Register Herald* (Beckley, WV), September 28, 2010. http://www.register-herald.com/disaster/x124778474/Upper-Big-Branch-mine-disaster-Special-online-section.

Urbina, Ian, and Michael Cooper. "Deaths at West Virginia Mine Raise Issues About Safety." *New York Times*, April 6, 2010. http://www.nytimes.com/2010/04/07/us/07westvirginia.html.

Ward, Ken, Jr. "MSHA Seeks Order to Close Massey Mine." *Charleston (WV) Gazette*, November 3, 2010. http://wvgazette.com/News/201011031294.

———. "State backs up previous Upper Big Branch reports." *Charleston (WV) Gazette*, February 23, 2012. http://wvgazette.com/News/201202230076.

Air Quality

Pope, C.A. "Respiratory Disease Associated with Community Air Pollution and a Steel Mill, Utah Valley." *American Journal of Public Health* 79, no. 5 (1989): 623–628. http://ajph.aphapublications.org/doi/abs/10.2105/AJPH.79.5.623.

Stradling, David. *Smokestacks and Progressives: Environmentalists, Engineers and Air Quality in America 1881–1951*. Baltimore: John Hopkins University Press, 1999.